MANAGING ADAPTATION TO CLIMATE RISK

Resilience has become the new buzzword – in government, health, energy and disaster management sectors. Climate change is the single largest threat to sustainable development, and addressing climate risk is a challenge for all.

This book calls for greater collaboration between climate communities and disaster development communities to tackle the challenges faced in addressing climate risk reduction. It evaluates approaches used by each community to reduce the adverse effects of climate change, and argues that adaptation focused on peoples' livelihoods, rather than technology, is the best way to reduce risk. One area that offers some promise for bringing together these communities is through the concept of resilience: a process that embeds capacity to respond to and cope with disruptive events. The book explores various models of resilience and concludes by evaluating the scope for a holistic approach where these communities can build communities resilient to climate driven risks.

Managing Adaptation to Climate Risk provides a conceptual framework for future debate and will be of interest to students and researchers in Development Studies, Environmental Studies and Disaster Risk Reduction.

Geoff O'Brien is Senior Lecturer in the Department of Geography at Northumbria University, UK.

Phil O'Keefe is Professor of Environmental Management and Economic Development in the Department of Geography at Northumbria University, UK.

MANAGING ADAPTATION TO CLIMATE RISK

Beyond fragmented responses

Geoff O'Brien and Phil O'Keefe

LONDON AND NEW YORK

First published 2014
by Routledge
2 Park Square, Milton Park, Abingdon, Oxon OX14 4RN

Simultaneously published in the USA and Canada
by Routledge
711 Third Avenue, New York, NY 10017

Routledge is an imprint of the Taylor & Francis Group, an informa business

© 2014 Geoff O'Brien and Phil O'Keefe

The right of Geoff O'Brien and Phil O'Keefe to be identified as authors of this work has been asserted by them in accordance with sections 77 and 78 of the Copyright, Designs and Patents Act 1988.

All rights reserved. No part of this book may be reprinted or reproduced or utilised in any form or by any electronic, mechanical, or other means, now known or hereafter invented, including photocopying and recording, or in any information storage or retrieval system, without permission in writing from the publishers.

Trademark notice: Product or corporate names may be trademarks or registered trademarks, and are used only for identification and explanation without intent to infringe.

British Library Cataloguing in Publication Data
A catalogue record for this book is available from the British Library

Library of Congress Cataloging-in-Publication Data
O'Brien, Geoff.
Managing adaptation to climate risk: beyond fragmented responses/
Geoff O'Brien and Phil O'Keefe.
  pages cm
  Includes bibliographical references and index.
  1. Human ecology. 2. Climatic extremes.
  3. Climatic changes–Effect of human beings on.
  I. O'Keefe, Philip. II. Title.
GF71.O27 2013
363.738'747–dc23   2012049015

ISBN: 978-0-415-60093-4 (hbk)
ISBN: 978-0-415-60094-1 (pbk)
ISBN: 978-0-203-83691-0 (ebk)

Typeset in Bembo
by Sunrise Setting Ltd, Paignton, UK

Printed and bound by CPI Group (UK) Ltd, Croydon, CR0 4YY

CONTENTS

	List of figures	*vi*
	List of tables	*viii*
	List of boxes	*ix*
	Acknowledgements	*x*
	Abbreviations and acronyms	*xi*
	Introduction	1
1	The sustainability problem	8
2	The climate journey	31
3	Climate extremes: does a post-normal approach make sense?	51
4	Disaster management	72
5	Adaptation	94
6	The concept of resilience	118
7	Development	148
8	Social capital and social learning	163
9	Conclusion	182
	Notes	*189*
	References	*191*
	Index	*212*

FIGURES

1.1	Mapping sustainable development	13
1.2	World energy consumption, 1990–2035 (quadrillion Btu)	16
1.3	The carbon emissions of the world's top 10 emitters in 2009	17
1.4	Tonnes of CO_2 per capita in 2009	18
1.5	Energy ladder	22
1.6	Conceptual visualisation of a resilient energy system	23
1.7	Conventional and distributed energy systems	24
1.8	Shifting the direction	24
3.1	Temperature increase projection	52
3.2	Post-normal science	61
3.3	Post-normal risk management	67
3.4	Flexible adaptation and mitigation pathways	69
4.1	Human–environment interactions	74
4.2	Disaster management cycle	75
4.3	Conventional disaster management within human–environment interactions	89
5.1	The adaptation continuum	95
5.2	Temporal dimensions of the adaptation continuum	99
5.3	The DFID sustainable livelihoods framework	103
5.4	Post-normal risk management	107
5.5	Heptagon of the determinants of adaptive capacity	110
5.6	Octagon of willingness	113
5.7	Adaptation uncertainty	114
5.8	Conceptual framework for a scenario-neutral approach to adaptation planning	115

6.1	Three-dimensional stability landscape with two basins of attraction showing, in one basin, the current position of the system and three aspects of resilience	121
6.2	Changes in the stability landscape have resulted in a contraction of the basin the system was in and an expansion of the alternate basin. Without itself changing, the system has changed basins	122
6.3	Representation of four systems and flows between them	123
6.4	Resilience: another dimension of the adaptive cycle	124
6.5	Connections between three levels of panarchy	126
6.6	Conceptual linkages between vulnerability, resilience and adaptive capacity	132
6.7	Enabling framework for resilience building	136
6.8	Resilience as bouncing forward	138
6.9	From sectoralism to holism	140
7.1	Structure of the Sphere Handbook Humanitarian Charter and Minimum Standards in Humanitarian Response	158
7.2	The disaster management cycle	159
7.3	The disaster-development management cycle	160
8.1	Frequency of references to social capital in the social science citation index, 1990–2010	165
8.2	Conceptual framework for the social capital assessment tool	167
8.3	Conceptual framework for measurement and analysis of social capital of the 'policy research initiative'	168
8.4	Most common measurement models of social capital	172
8.5	Learning processes	174
8.6	New learning cycle	180
8.7	Learning for a low carbon pathway	181
9.1	Mapping the resilience debate	187

TABLES

1.1	Schools of sustainability	12
1.2	Renewable energy share of global final energy consumption in 2009	15
1.3	Conference of the Parties timeline	19
2.1	Summary of adaptation funding under UNFCCC	33
4.1	Technocratic model of disaster management	79
4.2	Timeline of events	80
4.3	Contrasting characteristics of the traditional view of natural hazards and the new perspective presented by climate change	90
4.4	Definitions of vulnerability	91
5.1	Changes needed for a new resilience paradigm	98
5.2	The adaptation continuum framework features	101
5.3	Determinants of adaptive capacity	111
6.1	Resilience concepts	129
6.2	Characterising disasters	133
9.1	Schools of resilience	183

BOXES

2.1	Summary of findings of The Copenhagen Diagnosis	39
3.1	Climate deniers	55
3.2	Floods and landslides in the Republic of Korea	67
4.1	The United Kingdom and BSE	78
6.1	Basins of attraction	120
6.2	Ten-point checklist – essentials for making cities resilient	144
7.1	Changes in the humanitarian sector in response to the Rwandan crisis	156
7.2	The Hyogo Framework for Action, 2005–2015: building the resilience of nations and communities to disasters	160

ACKNOWLEDGEMENTS

First, to staff at Routledge who put up with our promises of delivery and constantly had to reconfigure their schedules when we did not produce on time. Second, to our close colleagues, Bernard Manyena, Janaka Jayawickrama and Komal Aryal with whom we have debated these issues for the last decade. Third, we owe a great debt to Manuela Scharf who shaped the chapter on social capital and provided an editorial hand in the final production.

Most books have dedications. We just wish to acknowledge that our children are probably right when they suggest both fathers are mad.

Newcastle upon Tyne, November 2012

ABBREVIATIONS AND ACRONYMS

AA	Arctic Amplification
AFB	Adaptation Fund Board
ALNAP	Active Learning Network for Accountability and Performance
AOSIS	Alliance of Small Island States
AWG-KP	Ad Hoc Working Group on Further Commitments for Annex I Parties under the Kyoto Protocol
AWG-LCA	Ad Hoc Working Group on Long-term Cooperative Action under the Convention
BAP	Bali Action Plan
BAPA	Buenos Aires Plan of Action
BCM	Business Continuity Management
BRIC	Brazil, Russia, India and China
BRM	Bali Road Map
BSE	Bovine Spongiform Encephalopathy
Btu	British Thermal Unit
CBDM	Community-Based Disaster Management
CBRN	Chemical, Biological, Radiological and Nuclear
CCA	Climate Change Adaptation
CDM	Clean Development Mechanism
CJD	Creutzfeldt–Jakob Disease
CO_2	Carbon Dioxide
COP	Conference of the Parties (UNFCCC)
CRU	Climatic Research Unit
DFID	Department for International Development (UK)
DHS	Department of Homeland Security (US)
DRR	Disaster Risk Reduction
ECHO	European Community Humanitarian Office (formally: European Commission – Humanitarian Aid and Civil Protection)

ETS	Emissions Trading System (EU)	
EU	European Union	
FAO	Food and Agricultural Organisation	
FDI	Foreign Direct Investment	
FEMA	Federal Emergency Management Agency (US)	
FMD	Foot and Mouth Disease	
GATT	General Agreement on Tariffs and Trade	
GCF	Green Climate Fund	
GCM	Global Citizens Movement	
GDI	Gross Domestic Investment	
GDP	Gross Domestic Product	
GEF	Global Environment Facility	
GEMS	Global Environmental Monitoring System	
GHA	Global Humanitarian Assistance	
GHG	Greenhouse Gas	
GTI	Great Transition Initiative	
GVA	Gross Value Added	
HFA	Hyogo Framework for Action	
HRBA	Human Rights-Based Approach	
IBRD	International Bank of Reconstruction and Development	
ICLEI	International Council for Local Environmental Initiatives	
ICRC	International Committee of the Red Cross	
IDNDR	International Decade for Natural Disaster Reduction	
IEA	International Energy Agency	
IFRC	International Federation of Red Cross and Red Crescent Societies	
IMF	International Monetary Fund	
INGO	International Non-Governmental Organisation	
IPCC	Intergovernmental Panel on Climate Change	
JEEAR	Joint Evaluation of Emergency Assistance to Rwanda	
JI	Joint Implementation	
LDC	Least Developed Country	
LIBOR	London Inter-Bank Rate	
MDGs	Millennium Development Goals	
MOP	Meeting of the Parties (UNFCCC)	
MRV	Measurement, Reporting and Verification	
MSF	Médicins Sans Frontiéres	
NAMAs	Nationally Appropriate Mitigation Actions	
NAP	National Adaptation Plan	
NAPAs	National Adaptation Programmes of Action	
NCAP	Netherlands Climate Assistance Programme	
NEPIs	New Environmental Policy Instruments	
NGO	Non-Governmental Organisation	

NHS	National Health Service (UK)
OCHA	Office for the Coordination of Humanitarian Assistance
ODA	Overseas Development Aid
OECD	Organisation for Economic Co-operation and Development
PNS	Post-Normal Science
REDD	Reducing Emissions from Deforestation and Forest Degradation
RRF	Regional Resilience Forums
SBI	Subsidiary Body on Implementation
SBSTA	Subsidiary Body for Scientific and Technological Advice
SCAT	Social Capital Assessment Tool
SCCF	Special Climate Change Fund
SES	Socio-Ecological System
SIDS	Small Island Developing States
TEC	Tsunami Evaluation Coalition
TWC	The Widening Circle
UKCIP	United Kingdom Climate Impact Programme
UNCED	UN Conference on Environment and Development
UNDG	United Nations Development Group
UNDG-HRM	United Nations Development Group Human Rights Mainstreaming Mechanism
UNDP	United Nations Development Programme
UNEP	United Nations Environment Programme
UNFCCC	United Nations Framework Convention for Climate Change
UNFPA	United Nations Population Fund
UNHCR	United Nations High Commissioner for Refugees
UNICEF	United Nations Children Fund
UNISDR	United Nations International Strategy for Disaster Reduction
USAID	US Agency for International Development
VAR	Value Added Reseller
WCDR	World Conference on Disaster Reduction
WCED	World Commission on Environment and Development
WFP	World Food Programme
WMO	World Meteorological Organisation
WSSD	World Summit on Sustainable Development
WTO	World Trade Organisation
WWF	World Wildlife Fund

INTRODUCTION

> It's not the strongest of the species that survives, nor the most intelligent, but the one that's most adaptable to change.
> (Attributed to Charles Darwin)

The quote, attributed to Darwin, encapsulates the message of this book. In the face of what many regard as the greatest physical threat to our species, accelerated climate change and increased variability, it is increasingly evident that adapting to new climate conditions is essential if we are to survive and prosper. We cannot fix climate change. We may find solutions to reduce emission rates but the cumulative additions already loaded into the atmosphere will drive change in the climate system. We will have to adapt.

This book is part of our shared journey. We know where we want to go but cannot see, at the moment, how to get there, except to say that 'We need other people'. But we can see where we do not want to go.

We do not want to go back to a carbon-based energy economy, but there are many people and organisations that do. At the present moment, carbon is still the cheapest source of fuel even if the full environmental costs of it have not been calculated. It is the fuel with the global infrastructure and buried capital that still wants to earn a profit. It is also the fuel that increases the global greenhouse gas (GHG) burden and thus accelerates climate change. As we write, the Arctic this summer is increasingly free of ice – 70 per cent attributable to human activity, 30 per cent to natural processes of warming. This is not a sustainable future.

In many senses, our take on sustainability drives our take on resilience. While the drive to sustainable development has hit the buffers, not least because of the collapse of finance capital, there are still three interpretations of what it should mean. The first is ecological, where natural capital with its associated stock and yield must not capture more yield (off-take) at the expense of eroding the stock. The second is conventional economics, where capital, both stock and revenue, must not be consumed in such a way that revenue demands erode stock. The final version is that of an open-ended sustainability, one that seeks to build both nature and capital to establish new equilibrium rather than, as in the case of the first two, seeking to reconstruct past equilibrium.

We adopt a somewhat similar approach to resilience, the buzzword of the twenty-first century. The classical interpretations from ecology and conventional economics are to seek a resilience that has 'bounce-back' ability, a return to the status quo. We seek to build resilience as a process that allows bounce-forward ability, which creates new opportunity.

Resilience, we noted, is the buzzword. It originally cropped up around emergency planning, moving outward to disaster risk reduction (DRR) and climate change adaptation (CCA) efforts. Then it started creeping into local government language, even national government. Initially, we were not concerned about this, although we realised that when resilience became a goal of organisations themselves, they made themselves resilient, not the communities they served. Then came the current financial crisis and the themes of rights, responsibilities and resilience of which authorities spoke to communities, informing them that they had no rights and the authorities had no responsibilities and so the communities had better be resilient – a classic 'blame the victim' approach.

In exploring what kind of resilience we do not want, together with what we do want, we come upon many examples of good practice. What we do not have, however, is a universal model of good governance where horizontal decision structures are as important as vertical ones. In that sense, we, like everyone else, are just at the beginning. By avoiding some of the pitfalls outlined in this book, progress can linger longer.

People are adept at change. This ability is the reason why we have become the dominant species. Our ingenuity, flexibility and adaptability have helped us to respond to both changing conditions and new opportunities. Our history shows how remarkable the human species is in dealing with problems. Changing climate conditions, following the end of the Younger Dryas period some 11,500 years ago, helped the shift from hunter–gatherer to agriculturalists in response to new conditions and an expanding population. We needed new sources of food; agriculture allowed us to exploit new sources of food. This transformed lifestyles. Settlements grew into cities, which, in turn, demanded an expanded agriculture to support growing populations. Government, religion and laws codified our lives. The ability to communicate through both writing and trade introduced and spread new ideas. This had profound impacts on development patterns. The industrial revolution was a step change in how Western Europe viewed and used the natural world. It was the exploitation of fossil fuels that powered profound economic and social change while simultaneously setting the seeds of the problems we are now facing.

We now face a dilemma. Some see it as post-modernism – a sense of little local exploitation of possibilities coupled with a rejection of the meta-narrative of global economic growth. We see it as a planetary phase, where the meta-narrative is to address the challenge of global poverty alleviation while simultaneously adapting to accelerated environmental change generated by global warming. In both meta-narratives the dialectical tension is between the global drivers and the local responses. And behind these global–local dialectical tensions lies another – our global models cannot and will never predict local outcomes.

We have come a long way in a relatively short period of time. It is only recently that we have begun to understand the impacts of our history. The scale of the challenge has only just begun to emerge. We are locked into an energy system that is changing the world in ways that we do not yet fully understand, especially locally. We recognise that the stable environmental conditions that have allowed us to flourish and prosper are being altered by our actions. Will we be able to respond? Can we move to a low carbon pathway? Can we adjust quickly enough to new conditions? Can we adapt in a way that is fair to current and future generations? These are challenging questions.

The latter questions are the focus of this book. The reason for this is simple. The atmosphere already holds loadings of GHGs from past human activities. Post-Durban, there seems little prospect of a global agreement on reducing GHG emissions. That means that more GHGs will continue to be loaded into the atmosphere. There is almost universal agreement that GHG levels have increased since the pre-industrial era. The reasons for this are directly related to human activities, principally fossil fuel use and activities such as deforestation and land use changes. There is almost universal agreement that additional GHGs will accelerate global warming. The areas of disagreement centre on the implications of our actions. Although there is general agreement that the atmosphere will warm, there is strong disagreement on the outcome.

How much the atmosphere will warm and what the likely effects on the climate will be are contentious issues. The Intergovernmental Panel on Climate Change (IPCC) suggests that by the end of the century the global mean average temperature could have increased by between 1.6 and 6 degrees Celsius. This would drive increased icecap and glacial melting, which would increase the average global sea level by between 28 and 43 centimetres. There is a general political consensus that the global average temperature increase should not exceed 2 degrees Celsius because a higher average temperature would lead to dangerous climate change. But the final average temperature outcome is dependent on which emission reduction pathways are achieved. With no agreement post-Durban other than an acknowledgement that we should aim to restrain the global average temperature increase to no more than 2 degrees Celsius, it is clear that human activities are producing a different and uncertain climate future. There is scientific concern that claims the IPCC is too conservative and the 2 degrees Celsius target is unrealistic and that we should expect a rise of at least 4 degrees Celsius in the average global temperature.

With a lack of political consensus and the prospect of one unlikely, coupled with uncertainty about how the climate will behave, humanity has entered an era of 'produced' unknowns. There are a number of areas of uncertainty. First, political uncertainty. With no agreement on reducing GHG emissions at Durban there are confused signals being sent to national governments about the timetable and the target. There is no agreed stabilisation target (GHG concentration) or timetable from which rates can be determined. The European Union (EU) has set a target and date aimed at keeping the increase in global average temperature at or below 2 degrees Celsius, but there is little clarity about which stabilisation target will achieve that. Will a GHG concentration of say 550 ppm (parts per million) achieve this? If so, can a political consensus be reached? The longer it takes to reach that consensus, the steeper the cuts that will be needed. Should it be a global commitment or should those countries most responsible for historic and present emissions carry the brunt and reduce their emission rates at a faster rate so that others can grow their economies? Evidence from the Kyoto experience suggests any kind of differentiated agreement that will deliver meaningful cuts will be problematic.

The second area of uncertainty is the science of climate change. There is almost universal agreement that human activity is contributing to the concentration of GHGs in the atmosphere, but there are uncertainties about the outcomes. There is a range of opinions about how much the global average temperature will increase and, although climate models are vastly improved, there is considerable argument about impact. The lack of political certainty about timescales and targets only adds to this.

The third area of uncertainty is what will happen as the temperature increases. The rise in temperature will generate more extreme weather events. There is a clear difference between what is meant by climate and weather. Climate is the average weather over a long-term period; weather is the prevailing conditions at a particular time. Changes to factors that determine climate conditions will lead to more uncertain weather patterns. As human actions contribute to the change of some of these factors, and thus accelerate changes in climate, this means that more extreme events are likely; however, it is the less extreme or mundane changes, such as increased drought, that will have serious consequences, particularly for vulnerable communities and ecosystems.

The fourth area of uncertainty relates to the prospect of dangerous climate change. The shift in climate conditions might trigger positive feedback. This is where increased temperatures might trigger other events that contribute to further warming, for example the melting of parts of the tundra releases methane, a potent GHG. This could accelerate climate change to a point where any actions by humans to reduce the production of greenhouse gas would be swamped by other processes. Dangerous climate change is where change is so rapid that natural systems do not have time to adapt, risking the prospect of large-scale extinction. Many of the ecosystem services on which humanity depends could be lost.

Fifth are the conditions that would prevail if humanity was to successfully control its activities and stabilise GHG concentrations. What would be the prevailing climate conditions and, for example, how would this impact the distribution of precipitation and weather patterns? There would be considerable implications for agriculture as well as human habitats and ecosystems.

Sixth are concerns about what would happen next. Stabilising the concentration of greenhouse gas means that the use of fossil fuels would be severely curtailed. Over time the natural carbon cycle would start to lower the concentration level and the climate would start to cool. Would there be a return to climate conditions that prevailed prior to the large-scale use of fossil fuels and how long would this take?

This is a lengthy way of saying that the future, from a climate and weather perspective, is very uncertain. In the run-up to and aftermath of Durban, there is a growing consensus of the need to adapt to new but unknown (and unknowable) conditions, as well as to cope with and respond to ongoing disruptions. This means dealing with the current risks generated by human interference with the climate system and planning for change in the medium to longer term. This is both current and ongoing risk reduction.

The overarching context for this is sustainable development because this recognises the need to do things today in a way that lowers risk both now and in the future. From a global institutional perspective there are three communities that need to collaborate to ensure effective adaptation and risk reduction. The first is the climate community under the UNFCCC (United Nations Framework Convention for Climate Change), which has expertise on impacts. The second is the disaster management community under the UNISDR (United Nations International Strategy for Disaster Reduction), which has expertise in disaster risk reduction. The third is the development community under the UNDP (United Nations Development Programme) in terms of the developing world, and under national governments in a more general sense, which has expertise in sustainable development. These are the three communities (3Cs) of climate change that will need to co-create solutions with actual communities around the world. That collaboration must, however, be led by the UNDP so that it is people centred.

Adaptation has long been argued as a major issue for poorer nations. That is still the case. More recently it has been recognised that adaptation is an urgent need for all nations. What is important to note is that there is no simple formula or strategy for adaptation. There is no definitive definition. The IPCC does define adaptation in terms of climate change, but so do others. What is clear is that there is a huge learning agenda from individual, community and institutional levels about what is meant by effective adaptation. Everyone will need to learn and share information. This will depend on the collective ownership of the problem and recognition that there is no single solution. Learning will need to be dynamic and flexible and operated through multiple pathways under a common ethical framework. There will be a need for a common language, common definitions and tools, as well as understanding of strengths and weaknesses, for example whether loss will be defined in terms of human lives and livelihoods or economic damage.

Where next?

Climate change is the single largest threat to the attainment of the Millennium Development Goals (MDGs) and sustainable development. Addressing climate risk is a challenge for all. Although risk management is an established practice, adaptation as a strategy for managing climate will require new approaches. Conventional risk management is premised on a systematic review of risks, where judgements about severity or likelihood are based on a number of factors, such as previous experience (e.g. young male drivers attract a higher insurance premium because insurance data show they are more likely to be involved in road accidents compared to other road users) or modelling techniques that evaluate the potential for failure that could lead to a disaster. This form of risk management can be characterised as experiential because it draws upon existing knowledge of risk. The new risks that are introduced, for example through technological innovation, are likely to be similar to existing risks because many innovations tend to build on existing developments – rarely does innovation introduce a step change that has little or no connection with existing developments.

Climate change and variability is a different challenge because there are many uncertainties and little or no experience on which to draw. The one certainty is that climate conditions will change, but the speed and distribution of impacts is unknown. The impacts can be characterised because they will be weather related – drought, flood, storm and so on – but, given the global nature of the changes, predicting the where, when and severity will be problematic. Uneven development patterns across the globe will, at least, give some indication of vulnerable populations and that, coupled with informed projections of likely weather patterns, for example more severe storms in storm prone areas, will show the nature and scale of problems to be addressed. Although it is very likely that there will be some spectacular events, such as Hurricane Katrina, that will grab the attention of the media, there will be many more mundane and largely unreported events that will severely erode livelihoods and claim lives, particularly those of the vulnerable.

Dealing with such events typically falls under the aegis of disaster management. The response to disruptive events is predicated on risk assessment ('what is likely to go wrong') and a reactive response when something does go wrong. Post-disaster recovery is usually the responsibility of other agencies. For routine events the reactive approach can be effective, but in the event of multiple and simultaneous disruptions the capacity of the response function could be quickly overwhelmed and exhausted. In the recovery phase there is a danger that a lack of interaction between the response and recovery functions could lead to vulnerabilities being exacerbated and disaster events repeated.

It is clear that adaptation is needed and that careful thought must be given to the approach. Risk management as currently practised will have a role to play, but the most significant changes in approach must focus on how to prepare for a largely unknowable future. In short, the focus of all three institutional communities must be on preparedness, i.e. enhancing self-help and self-reliance. Effort must be directed at building capacity to respond to disruptive events. This is a process of building resilience from the level of the household through to the community and then to the nation state and, arguably, the market. Many of the responses may be straightforward, for example the use of drought-resistant crops where rainfall agriculture is jeopardised by variable precipitation patterns. Others may be more drastic and require wholesale relocation, for example in the event of inundation of low-lying coastal areas. Where relocation means crossing national borders, resilience building at the international level is needed. Global and regional institutions will have to rapidly develop adaptive capacity to cope with such complex issues.

At the very least, nations throughout the world will need to start planning. This will require the active collaboration between the climate, disaster and development communities, particularly in response to complex challenges, such as failed or contested states, cross-boundary impacts and issues, and the role of the private sector in creating and reducing vulnerability. Although these three bodies are located in the UN, each has its own drivers. UNFCCC is impact focused but has access to significant funding streams. UNISDR has considerable expertise in disaster risk reduction and has recently shifted from a reactive to a proactive stance but lacks financial capacity. The development community is focused on poverty alleviation, notably the MDGs, but relies on donors and is shaped by short-term priorities because donors are generally results focused. All three bodies are committed to finding solutions to the problems they face within a sustainable development context. In that sense, the 3Cs share a commonality of purpose. The challenge is to build that commonality in a way that practically engages and learns 'with' the actual communities who affect and are affected by our attempts to respond to climate change.

This book will use the concept of resilience as the vehicle for exploring how to build that commonality. The resilience concept has been used in a number of disciplines. The concept used here is that of the coupled human–environment system. Human actions can disturb the surrounding environment either positively or negatively. Similarly, environmental change can disturb human systems, such as agriculture, upon which people depend. The starting point is thinking about preparedness. In climate terms this means using the latest scientific techniques to evaluate the types of threats that are plausible and developing appropriate information exchange systems, working with financial institutions, and effectively utilising existing and new funding streams for adaptation. In disaster management terms this means a greater focus on resilience building to enhance adaptive capacity, a greater emphasis on pre-event planning and a greater focus on community-based approaches. In development terms this means a greater emphasis on climate proofing in conjunction with both the climate and disaster management function. This can be likened to building a continuum between these communities that has the overall aim of building community resilience. With a very uncertain future, our aim is to be as prepared as possible and the most effective way of doing that is through a coherent approach that allows the climate, disaster management and development communities to work and engage with actual communities more effectively. These are the themes explored in this book.

1
THE SUSTAINABILITY PROBLEM

> The difficulty lies, not in the new ideas, but in escaping the old ones, which ramify, for those brought up as most of us have been, into every corner of our minds.
> (John Maynard Keynes (1883–1946))

Introduction

Resilience is the buzzword. From local government to climate change, resilience drips everywhere. It emphasises the ability of people to adapt to changing, often threatening situations. Adaptation – thoughtful, planned action to make a world that is more secure – is one of the unique characteristics of being human. So why are we concerned about the current use of resilience?

First, a negative. Politicians increasingly use the phrase 'resilience' to cover up a lack of intervention. Community resilience is simply another way of saying there is no investment except for what you generate. It is a big-time cop-out as politicians try and rescue the financial institutions that created so much uncertainty and hardship in the first place. It is a cover up for the 'big society' that will not happen without collective investment, an investment that must start with rich people paying for a social future in proportion to their wealth. At the heart of resilience is the recognition that adaptation can only be generated as a shared future if poverty alleviation is the key driver. As current neo-liberal politics generate poverty, building resilience requires overturning the current version of capitalism and political leadership that supports it. Do not allow current political leadership to claim the term resilience when it really means, in their words, 'We are not in this together'.

Second, because the material drive is for poverty alleviation, resilience debates have to be focused in and around both global and local initiatives. At a global level, the two key debates are about the delivery of the Millennium Development Goals (MDGs), essentially about the creation and sharing of wealth, and the need to address climate change as the dominant threat to the global commons. Both of these debates are couched in the language of sustainable development where, if Rio+20 is to be a guide as well as the failure to generate a post-Kyoto climate agreement, there is a demonstrated erosion of commitment to action. The absence of equity as the key to sustainability, and thus building resilience, is the horrifying silence of international negotiations conducted by nation states.

Third, it is this absence that takes us back to unpacking the climate change negotiations to see precisely what went wrong. The science is robust and peer reviewed despite naysayers. The popular take ('where is the global warming?') is a misdirection. The science indicates an acceleration of global warming (i.e. there is already a period of warming, only it is getting faster) because of increased greenhouse gas emissions. These increased greenhouse gases are associated with gases, particularly carbon, that are generated in modern economies. To address the threat, resilience capacity must allow for mitigation and adaptation technologies to stop us harming ourselves and harming nature.

Fourth, even where we are making interventions, they continue to be self-serving. Mitigation, implying the creation of new technologies, significantly overshadows adaptation, making livelihoods less resilient. In many ways this is no surprise because it is the developed countries that dominate the discussion and financing of climate change, creating a twenty-first century bio-imperialism. But as environmental threat intermingles with financial catastrophe, the hope of a renewables, energy-efficient future fades against a background of increasingly dirty fuels, including tar sands and expanding exploration into pristine environments such as the Arctic. Adaptation is mere lip service for developed countries.

Fifth, even if mitigation were successful, it is essentially a subset of adaptation. Within both the development debate and the climate change debate, adaptation is seen as a major challenge for developing countries. The irony is a political economy of adaptation where those who are least guilty of carbon crime pay the penalty of its impact. Resilience building is denied and instead local communities are urged by the developed world to build disaster risk reduction strategies. There are two problems, however, namely misunderstanding the nature of those at risk and ensuring the response system is not centrally controlled. The misunderstanding of those at risk begins by stating that risk = vulnerability × hazard. This is an unworkable equation, not simply because it cannot be transposed, but because a vulnerability argument is one that understands that the nature of everyday life both constructs risk but simultaneously ranks those risks lower than other problems of daily life. Putting hazard into the equation simply repeats the limitations of traditional disaster paradigms. Moreover, there is clear evidence that political structures can only do top-down interventions because the people cannot be trusted. Despite the language of empowerment, there is none – except to say 'Make yourselves resilient', and *sotto voce*, 'Without resources from us'.

Sixth, we argue that the problem essentially originates in a misunderstanding of development. The material evidence that strikes us is that China has moved 300 million people out of poverty in 10 years, an unprecedented feat at the start of the twenty-first century. It has successfully done so without a neo-liberal approach to development, while those who pursue such politics stagnate or go backwards. For at least a generation the focus of development has been the delivery of sustainability where three competing interpretations vie for dominance. Is sustainability of the environment, the market or the people to be the central thrust of sustainable development? The first two interpretations argue from equilibrium, respecting only the status quo; the third suggests no equilibrium, only evolution – rapid evolution, even revolution. That is our preferred interpretation of sustainable development within an understanding of human rights. We find, however, very similar underpinnings to the three interpretations of resilience: two favour equilibrium and leave things as they are, while the third seeks evolution, progress. Mapping institutions suggests we have limited governance structures for both sustainability and resilience efforts.

Finally, in seeking progress, we put people first. Centrally, we wish to achieve a common wealth of entitlements and to ensure that, it means we have to build social capital in governance structures we trust. It requires a model of learning, not just from individual mistakes but also from lost opportunities that are frequently lost lives and livelihoods. It requires meaningful structures of governance beyond the emptiness of elections where on a day-to-day basis people's demands are heard. And we know what the demand is: a right to common wealth and a rejection of the erosion of entitlements and thus resilience.

The demands are openly socialist in intention, but these demands are built on the science available at the beginning of the twenty-first century and indicates that current capitalist regimes are doomed to failure, not just the regimes but the people they control. It is exciting to have a challenge for change, initially by exploring resilience and then beyond.

We start this book by looking at the climate change problem. This is a logical place as the climate problem sheds light on how we deal with problems or, in this case, how we have failed to deal with the problem. Subsequent chapters will address related issues, but first a brief discussion of where we are today.

Climate change

Accelerated climate change and increasing variability is the single greatest threat to the international goals of sustainable development and poverty reduction. It is also the first real challenge to the Western or developed world way of life. Lifestyles in the developed world are predicated on unfettered access to natural resources, in particular fossil fuels. In reality, our lifestyles are determined and defined by access to fuels and the types of energy services we demand (Giampietro et al. 2001; Bulkeley et al. 2011). The more energy we can access, the 'wealthier' we are. Compare the access that many have to energy resources in the developing world. Women in India regularly spend 2 to 7 hours each day collecting fuel for cooking, and in rural sub-Saharan Africa, many women carry 20 kilograms of wood fuel an average of 5 kilometres every day (O'Brien et al. 2007). In the developed world, access to personal transport, health, education and leisure services, all of which are high energy-users, are almost regarded as a right. Our homes have high comfort levels, and foreign travel for leisure purposes is the norm as opposed to the exception. Our consumerist lifestyle is highly dependent on access to energy resources.

The problem is quite simple. Fossil fuel use is increasing the amount of greenhouse gas in the atmosphere. GHGs act to trap the emitted radiation that is produced by the incidence of the Sun's electromagnetic energy on the Earth, which then warms the atmosphere. Without this greenhouse effect, the average temperature of the Earth would be about −20 degrees Celsius. From our perspective, Earth would be uninhabitable. The quantity of greenhouse gas in the atmosphere has a 'Goldilocks' effect: too little and the planet cools, too much and it warms, and just the right amount provides the temperature conditions for life to flourish. Current scientific thinking suggests that a concentration of greenhouse gas that existed prior to the industrial age, about 280 ppm, is just about right. At present, the concentration is about 390 ppm and rising at a rate of 1–2 ppm per year, and it appears highly likely that we will continue to add to GHG loadings.

Climate change is not new. The climate of the Earth has changed considerably throughout its evolution. Since life on Earth evolved, a number of factors have influenced the climate such as impacts of objects from space or increased volcanic activity; however, the most consistent influence has been the rotation of the Earth and its orbit around the Sun. This pattern has had variations. At times, the Earth moves further from the Sun and hence receives less energy, which tends to have a cooling effect. This pattern of change has been mapped by Malinkovitch and shows a strong correlation with climate episodes in the Earth's past (Dessler 2012).

The problem, in a nutshell, is that we produce too much GHG. The solution appears to be straightforward: reduce the amount of GHG we produce. But this is proving to be an almost impossible challenge. The UNFCCC was agreed to develop an international agreement to reduce GHG emissions. It also recognised the need to adapt to changes that are being driven by GHG already loaded into the atmosphere. But, as described in Chapter 2, the process of reaching any sort of agreement has been riddled with problems.

Dealing with the consequences of climate change is not just the responsibility of the Convention – governments throughout the world have disaster management organisations. Usually these address all types of hazards: technological, seismic and volcanic eruptions as well as weather related. Weather-related hazards tend to be the most frequent, for example storms, floods and droughts. It would seem sensible that there was collaboration because the frequency of weather-related events is beginning to increase. Of course, all of this has to happen within a sustainable development context – a complicated landscape of public and private actors with different and sometimes opposing governance priorities. This raises and partially answers the question of why so little progress is being made.

Is sustainable development part of the problem?

The concept of sustainable development was popularised by the World Commission on Environment and Development (WCED) (more usually called the Brundtland Commission) in its report *Our Common Future* (United Nations 1987). Brundtland posited that sustainable development was an intergenerational concept where meeting needs should not jeopardise opportunities, both now and in the future. Brundtland argues that the environment should not be seen as separate from either political or economic goals and that we can better understand the environment in relation to development, and development in relation to the environment. In short, the environment is where we live (and we cannot live without it) and development is what people do to enhance livelihoods.

The concept of sustainable development has been adopted internationally by groups of nations such as the EU, and by nations, cities, corporations, campaigners and communities. There is some ambiguity in the term sustainable development. As Redclift (2005) points out, what are we trying to sustain? Why are we doing so and for whom? The very woolly nature of the term allows a plethora of interpretations. Within the academic community, sustainable development is viewed from a number of perspectives, as shown in Table 1.1.

Ecological sustainability has a strong sense of the importance of the environment and the interconnected nature of differing ecological systems – and people are one

TABLE 1.1 Schools of sustainability

Subject matter	Ecological sustainability	Sustainable growth	Sustainable development
Major concerns	Ecosystems and biosphere	Economy, markets and prices	People's livelihoods, economy, society
Major disciplines, theoretical base	Natural sciences, biology, ecology	Neo-classical and new institutional economics, new political economy, rational choice theory	Agricultural and social sciences, 'old' and some 'new' economics, anthropology, ethnology
Basic world view	Equilibrium focused, nature centred	Equilibrium focused, anthropocentrist	Basically evolution focused, anthropocentrist
Major concerns with respect to quantification of sustainability	Rates of population growth, environmental degradation, loss of biodiversity, desertification, pollution, etc.	Rates of growth of income or consumption based on national accounts, market-valued flows of goods and services	Specific and aggregate social indicators, case studies of livelihoods, coping and conflict-solving strategies
Major policy prescriptions	Protect nature, educate people	Develop markets and internalise externalities	Empower people, develop institutions
Major goals	Ecological viability	Economic efficiency	Social efficiency, justice

Source: Adapted from Hatzius (1996).

of those systems. Sustainable growth is an area where many governments believe that they have achieved. In fact, the political rhetoric does not match reality, such as over-consumption, resource depletion, pollution and environmental degradation coupled to increasing poverty and growing inequity. This view has a strong focus on institutional capacity, which can deal with problems we face, and the power of the market to develop the most cost-effective ways of finding solutions to these problems. The evidence to date is not compelling. The sustainable development viewpoint is much more people-centred with its focus on livelihoods and localism. Although people are part of the problem, they are also part of the solution.

These perspectives are useful for analysing different discourses on sustainable development, but the reality is that people are political creatures and the dominant discourse on sustainable development is political. The impact of this is illustrated in Figure 1.1.

Figure 1.1 groups these discourses on sustainable development into three broad areas: status quo, reform and transformation. The position of the Organisation for Economic Co-operation and Development (OECD) demonstrates the problem. All members subscribe to a shared vision of multi-party democracy, capitalism and market-based economies. Member states, in general, are consumerist societies that enjoy high standards of living. Their energy economies are highly dependent on fossil fuels, but there is little indication of a major shift to a low carbon economy within the OECD. The Reform area includes those who believe that we need to

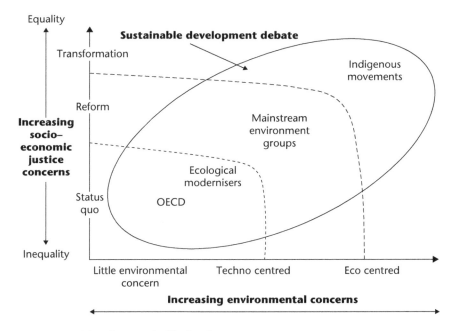

FIGURE 1.1 Mapping sustainable development

Source: Adapted from Hopwood et al. (2005).

reform the way our society functions, for example use of fiscal measures to support good environmental practice or to penalise adverse environmental practice. There are examples of nations that have adopted some of these ideas, for example the landfill tax in the United Kingdom and the use of the Feed-in-Tariff to support the use of renewable energy sources in Germany. In general, those advocating reform tend to be regarded as being outside of the political mainstream, particularly in the OECD. The Transformation area represents those who believe that existing structures within many countries are part of the problem, for example neo-liberal economics, which encourages consumerism that is responsible for environmental degradation (Sustainable Development Commission (SDC) 2009).

To simplify the point, there are two discernible trends in the interpretation of sustainable development. The first is a weak interpretation that is people focused, where people are perceived to be wholly separate from nature with a right to exploit nature. The weak interpretation of sustainability argues that technology and/or human ingenuity can solve the problems we have created. The second is a strong interpretation that is environment focused and recognises the need to live within ecological limits. In short, the strong view sees a completely different relationship with nature, where nature has rights, even if ultimately this is not for the benefit of humans. This view argues that the pursuit of materialism is heading in the wrong direction. As opposed to being a means to meet well-being, it is now an end in itself (Williams and Millington 2004; SDC 2009). The weak interpretation is very much a feature of the OECD and other nations, such as China, that pursue economic growth without respect for ecological limits. There is a need to redefine wealth as well-being as opposed to the acquisition of goods (SDC 2009). In Figure 1.1, status quo represents those who have a weak interpretation and, broadly, those within the Reform and Transformation areas would have a stronger interpretation.

The energy and mitigation problem

The Kyoto Protocol was a defining moment in the climate discourse. It was a triumph for those that subscribe to a weak interpretation of sustainability and advocate that markets are the best vehicles for producing cost-effective solutions for mitigating GHG emissions. It was also a triumph for the ecological modernisers who subscribe to the view that market measures, coupled to an acceptance that science is a force for the good, along with different political approaches and raised citizen awareness will deliver effective mitigation solutions. Their approach is iterative but remains within the bounds of the status quo, as it fails to advocate any reformative or transformative approaches to the energy system, the main producer of GHGs. The energy system has to adapt to new energy resources, such as solar power, and this requires a very different approach to system development. This very narrow technological viewpoint also fails to take into account other environmental problems or the impact of technological developments on ecological and human systems.

Nations such as Germany, which is placed within the status quo area, have made efforts to promote alternative energy technologies as part of a shift to a low carbon economy. Many cities throughout the developed world have adopted low carbon policies and many corporations have implemented environmental management systems or policies aimed at reducing resource use and lowering waste production. But the overriding aim within the status quo area is the pursuit of economic growth in order to remain globally competitive, so it is only as far as these nations will be prepared to go. In the midst of the current economic downturn, the clamour for more economic growth to produce employment is ever louder. There are few signs that any member of the OECD and other developed and rapidly developing countries will radically change course in the foreseeable future (International Energy Agency (IEA) 2011a). And even if we do, it will be difficult to adequately address the issue of cumulative GHG emissions (Bowerman et al. 2011).

The scale of the shift to a low carbon economy is huge. Table 1.2 shows the deployment of renewable technologies at the global level. Renewables account for 16 per cent of global energy consumption, but of that 16 per cent, 13.4 per cent is accounted for by traditional biomass and hydropower. Hydropower is a mature – if controversial – technology and the new energy technologies that produce heat and power (wind, solar, biomass, geothermal and biofuels) only account for 2.8 per cent of global energy consumption. This is a very small proportion of global energy needs and gives some idea of the scale of transformation of the energy system that will be needed if we are to realise a low carbon future.

The global population is predicted to increase to some nine billion people by 2050. The demand for energy will grow (IEA 2011a). Within the OECD, population growth is forecast to rise at a much lower rate than in the rest of the world. As the population rate continues to rise, the demand for energy services will increase, but the problem of energy poverty, both in developed and developing worlds, will not go away. The MDGs are supposed to address energy poverty in a developing world context. In the developed world, energy poverty is the responsibility of the state. But in a future world, the scramble for ever-diminishing energy resources is likely to generate sharp fluctuations in energy prices. It is likely that those in energy poverty today will remain so (O'Keefe et al. 2010). Figure 1.2 shows that the areas of major growth in energy use will be non-OECD countries, such as China, and it

TABLE 1.2 Renewable energy share of global final energy consumption in 2009

Source	Percentage of global energy
Fossil fuels	81
Nuclear	2.8
Traditional biomass	10
Hydropower	3.4
Biomass/solar/geothermal hot water/heating	1.5
Biofuels	0.6
Wind/solar/biomass/geothermal power generation	0.7

Source: Adapted from REN21 (2011).

16 The sustainability problem

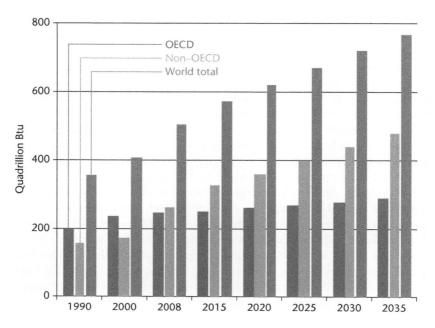

FIGURE 1.2 World energy consumption, 1990–2035 (quadrillion Btu)
Source: EIA (2011).

is very likely that the increase in demand will be met by fossil fuels. To contextualise this, it is worth highlighting the fact that although China is now the world's leading GHG producer, the United States, on a per capita basis, is still by far the leading producer of carbon emissions (see Figures 1.2, 1.3 and 1.4).

If fossil fuel use continues to rise, the concentration of GHGs will also rise. It is possible that the climate system could reach a 'tipping point' where irreversible and dangerous shifts in the climate system could occur (Anderson and Bows 2011; Bowerman *et al.* 2011; IEA 2011b). Abrupt changes could make conditions for human life very difficult. Lovelock envisages small groups of ragged survivors heading for the polar regions in search of habitable environments (Lovelock 2006). Lovelock does not envisage life on Earth becoming extinct, but he does think that human life could become extinct. Extinctions are not new. Our environment has evolved without human intervention. Our interventions could make conditions for human life untenable. Addressing climate change is an urgent need and we will have to adapt. UNFCCC is the international vehicle for addressing climate change. The best we can hope for is that some kind of agreement will emerge after 2020. This is political failure on a global scale.

The objective of UNFCCC is to stabilise GHG concentrations at a level that would prevent dangerous anthropogenic interference with the climate system (UNFCC 1992). Governance of UNFCCC is through annual meetings of the signatories to the Convention. This is known as the Conference of the Parties or COP

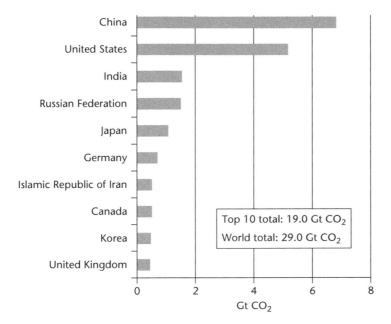

FIGURE 1.3 The carbon emissions of the world's top 10 emitters in 2009
Source: IEA (2011b: 11).

and an additional body was created in 2005 called Meeting of the Parties (MOP) of the Kyoto Protocol. COP meets annually to take stock of developments as well as to decide on future strategies for meeting the aims of the Convention and, more recently, the Kyoto Protocol – a listing of the annual meetings of the Convention is shown in Table 1.3. Climate change concerns focused strongly on reducing or mitigating GHG emissions. This gave a strong technology focus to discussions, a position that clearly shows the weak interpretation of sustainable development of the OECD nations.

It was clear by the Berlin COP (April 1996) that the aspirations of UNFCCC were not realisable. A combination of a fixed target and timetable did not recognise the differences between national energy economies, for example countries such as Norway, which produces most of its electricity by hydroelectric power, would have to load most of its cuts on motorists, a politically unacceptable position even in moderate Norway (Dunn 2002). At the Kyoto meeting, it was agreed to develop a protocol. The advantage of this was that signatories to the protocol would be required to meet emission targets that had been agreed by the signatories.

Where does this leave us? It seems that we can think of mitigation only as being solved through technology. This is not surprising because technology can be universalised, and many of the technologies and the means to produce them are held by those in the status quo camp. As Crist (2007: 33–34) argues, the framing

18 The sustainability problem

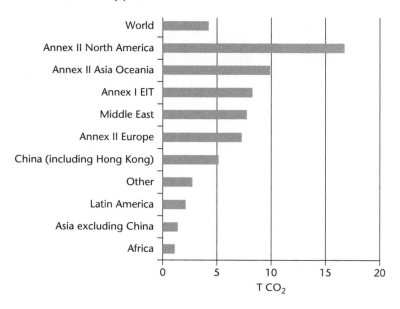

FIGURE 1.4 Tonnes of CO_2 per capita in 2009
Source: IEA (2011b: 12).

of climate change as a major problem, perhaps the biggest problem, implies the adoption of technical solutions only:

> Whether the call is for reviving nuclear power, boosting the installation of wind turbines, using a variety of renewable energy sources, increasing the efficiency of fossil-fuel use, developing carbon-sequestering technologies, or placing mirrors in space to deflect the sun's rays, the narrow character of such proposals is evident: confront the problem of greenhouse gas emissions by technologically phasing them out, superseding them, capturing them, or mitigating their heating effects. [...] Furthermore, if greenhouse gases were restricted successfully by means of technological shifts and innovations, the root cause of the ecological crisis as a whole would remain unaddressed. The destructive patterns of production, trade, extraction, land-use, waste proliferation, and consumption, coupled with population growth, would go unchallenged, continuing to run down the integrity, beauty, and biological richness of the Earth.

Shifting the energy trajectory

Globally we are very energy dependent. This places us in a bit of a quandary because we seem to use fossil fuels as if there is an unlimited supply and their use does not have any consequences. If we continue to use fossil fuels then, at some time in the not too distant future, we will run out. If we continue to emit GHGs at the current rate then we may reach a climate tipping point that may well see disruption across the planet. We need to make use of the plentiful supplies of non-polluting renewable energy resources.

This effort will have to transform the current energy system and avoid replicating the vulnerabilities and deficiencies of the current system. All countries need access to clean, affordable and reliable energy services that do not exacerbate climate change risks. In the developed world, access defines quality of life, while in the developing world lack of access constrains development. To make any progress towards achieving the MDGs requires a new approach to energy systems for poorer

TABLE 1.3 Conference of the Parties timeline

Date and location of COP	Summary of COP
COP 1 – April 1996 Berlin	Adopted the Berlin Mandate
COP 2 – July 1996 Geneva	Called for legally binding emissions targets
COP 3 – December 1997 Kyoto	Adopted the Kyoto Protocol that assigned legal emission reductions to those who ratify the Protocol. This will be discussed further.
COP 4 – November 1998 Buenos Aires	Adopted the Buenos Aires Plan of Action to help devise mechanisms to implement the Kyoto Protocol
COP 5 – October/November 1999 Bonn	Devoted to work on the adoptions of COP 4
COP 6 – November 2000 The Hague	Ended without any progress and was rescheduled
COP 6 – July 2001 Bonn	Adopted the Bonn Agreements
COP 7 – October/November 2001 Marrakech	Adopted the Marrakech Accords
COP 8 – October/November 2002 New Delhi	Adopted the Delhi Ministerial Declaration
COP 9 – December 2003 Milan	Agreed to use the Adaptation Fund to help developing countries to adapt to climate change
COP 10 – December 2004 Buenos Aires	Discussed the progress made in the last 10 years since the first COP
COP 11 – November/December 2005 Montreal	Entry of the Kyoto Protocol into force. The Montreal Action Plan was made as an agreement to extend the Kyoto Protocol beyond 2012.
COP 12 – November 2006 Nairobi	Improvements in support to developing countries and improvements in clean development mechanisms
COP 13 – December 2007 Bali	Adopted the Bali Action Plan to extend the Kyoto Protocol commitment period
COP 14 – December 2008 Poznan	Agreements made on a fund to help developing countries with the impacts of climate change
COP 15 – December 2009 Copenhagen	World leaders delayed making a decision on a climate change agreement and therefore did not achieve a binding agreement
COP 16 – November/December 2010 Cancún	Called for a Green Climate Fund to help developing countries; however, there were no further agreements on the Kyoto Protocol
COP 17 – November/December 2011 Durban	Agreed to have an agreement by 2020

nations. Equally, innovative thinking is needed to shape future energy policy in the developed and industrialising worlds.

Global discourses on energy futures have been primarily focused on mitigation and much of the discourse is supply-side focused. Energy system development is entering a supply-constrained era (Gupta *et al.* 2007). Geopolitical disruptions will add to existing system vulnerabilities. This places constraints on system development because security concerns require a shift to indigenous resources and climate concerns are driving a shift to low carbon and renewable resources. Supply-side issues are important but should not be viewed in isolation. The discourse should focus on re-thinking energy systems from both supply and demand perspectives. Future energy systems should contribute to sustainable development. There are many uncertainties for energy system development, such as price volatility driven by increased demand and a diminishing fossil fuel resource, geopolitical disturbances and the scale and timing of mitigation measures to avoid dangerous climate change. A resilience perspective can help shape system development so that it is able to more effectively respond to such disruptive challenges.

One of the major challenges for energy system development is energy poverty. In the developed world, energy poverty is still a problem (Boardman 2010). Conventional approaches to energy system development, for example large-scale interconnected electrical grid systems, have not and are unlikely to meet the needs of many people, particularly those of the poorest. Worldwide, nearly 2.4 billion people use traditional biomass fuels for cooking and nearly 1.6 billion people do not have access to electricity. By 2030, there is a risk that another 1.4 billion people will be in the same position (IEA 2003). Modi *et al.* (2006) argue the need to scale up the availability of affordable and sustainable energy systems. While there is little dissent from this view, this does raise questions about the pattern of future energy system development. The dimensions of the challenge for the energy system can be summarised as (O'Brien and O'Keefe 2006: 125):

- no adverse interference with the global climate system;
- wherever possible, using indigenous resources to minimise geo-political risks;
- appropriate to needs and long lasting;
- work within the context of the environment.

These dimensions apply globally. The scale of the challenge is huge and technologically sophisticated nations have experienced difficulties just in meeting climate obligations of the Kyoto Protocol. For example in the EU, initial assumptions that energy consumption could be reduced proved to be unfounded. Although some emission reductions were made, these were achieved through fuel substitution (in the United Kingdom, the 'Dash for Gas' that replaced coal fired stations) and structural changes (following the unification of Germany, many inefficient energy intensive industries were closed in the former East Germany) as opposed to real reductions in demand (Dunn 2002). These reductions were quickly surpassed by growth in the domestic and transport sectors, reflecting significant lifestyle changes. The drive for greater efficiency in generation, transmission, distribution and end-use has also been offset by demand growth (O'Keefe et al. 2010).

The Kyoto mechanisms, established to encourage technology transfer to the industrialising and developing world to aid development of sustainable energy systems, have instead become the developed world's vehicle for meeting the modest reduction targets of the Protocol (O'Brien et al. 2007). Part of the problem is the design of the mechanisms, for example the Clean Development Mechanism (CDM) provides monetary incentives for mitigation but none for sustainable energy system development in the host (Ellis et al. 2007). This form of mechanism for technology transfer is unlikely to effectively address energy poverty in the developing world. More broadly, the use of 'Cap and Trade' as a mechanism for reducing GHG emissions and stimulating technological innovations have shown little sign of delivering meaningful change (O'Keefe et al. 2010).

Addressing energy poverty requires a specific focus and one that is based on needs. Addressing sustainable development goals requires the use of low carbon and renewable resources and system development that enhances capacity. The development of energy systems is described as moving up the energy ladder, where the introduction of new fuels and technologies can be seen as moving to the next rung on the ladder. Simply progressing up the existing energy ladder will not address energy poverty effectively, as shown by Ladder 1 in Figure 1.5, which is predicated on the dominant model of progression, i.e. a system based on fossil fuels.

A new trajectory that focuses on the development of autonomous systems using indigenous resources, Ladder 2, offers the opportunity to enhance capacity. The Step Change shown in Figure 1.5 assumes the introduction of renewable technologies. This is not a simple technology transfer. The starting point is an assessment of needs, resources and availability of support systems before an entry point can be defined. This is a negotiated process where the technology producers would not dominate the specification process. Each step up the energy ladder will enhance capacity of the users. This embeds resilience. Resilience is a function of resources and capacity. Resources are defined as the livelihood capitals (human, social, natural, physical and financial). Sustainable energy systems are predicated on the availability of natural resources. Enhancing the capacity to operate, maintain and improve the system over time is a social learning process. It is also an institutional learning process that requires technology producers and support organisations to approach energy poverty and energy system development differently (O'Brien and Hope 2010).

Conceptually a resilient energy system brings together two actor groups, broadly those that own and use energy-producing technologies and those that develop and deploy those technologies. The starting point for elaboration is the household. From a technology perspective, end-use efficiency is vital. Capturing intermittent and diffuse renewable resources is a considerable challenge and it makes little sense to use them inefficiently. Existing technologies can deliver buildings that require virtually no space heating. This, along with embedded and localised systems, such

22 The sustainability problem

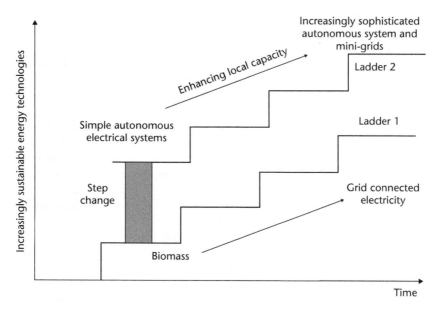

FIGURE 1.5 Energy ladder
Source: O'Brien et al. (2007).

as community-owned renewable technologies to supplement embedded capacity, can supply the required energy services. There are demonstration examples of autonomous or off-grid developments. Intervention is needed to improve standards for new builds and refurbishment. Similarly, higher standards are needed for fast turnover products such as household appliances.

When the user is able to interact with energy capture, and use and manage resources to meet needs, then it can be argued that this is a resilient energy system, i.e. one where human adaptive capacity is able to use indigenous resources to meet needs. Autonomy of this kind eliminates many of the vulnerabilities discussed earlier. This approach, shown in Figure 1.6, relies on technological development aimed at capturing and using indigenous renewable resources effectively and the willingness of the user to manage these resources within recognised constraints.

Although energy is vital to every aspect of human endeavour, the household represents a key part of human culture where many energy services are needed, such as heating, lighting, hot water and cooking. In Figure 1.6, energy resources are captured and/or stored either with embedded or localised technologies. The user interface provides information that enables the user to balance the energy service requirement to either available or stored resources. The technological challenge is to design a system that produces, over a period of time, more energy than is needed. Storage provides a buffer against source variability. Resilience-thinking implies some degree of over-design to cope with variability.

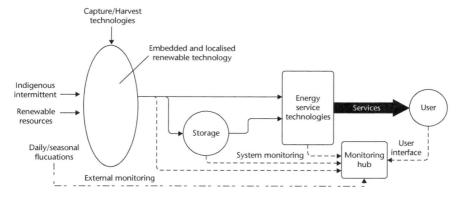

FIGURE 1.6 Conceptual visualisation of a resilient energy system
Source: O'Brien and Hope (2010).

The starting point for building resilience within new and existing stock is conceptualising building-technology combinations that will maximise resource capture. Many combinations are possible (autonomous, interconnected at street or area level and so on), which will begin to impact system architecture, for example in the electricity sector this means a shift from concentrated ownership of generation and distribution capacity and a model with many passive consumers to a more democratic model with many stakeholders, as shown in Figure 1.7. Besides addressing a number of systemic vulnerabilities of current structures, such changes can generate space for new forms of ownership and governance. This does not advocate a single model but argues that the principal driver for resilience building, predicated on entitlements and governance, is learning.

The dominant energy model is technically complex and capital intensive and has inherent technical vulnerabilities (Perrow 1999; Lovins and Lovins 1982). These are being compounded by geopolitical uncertainties of security of supply and more recently by instrumental threats (O'Brien 2009). Renewable resources are diffuse and intermittent and usually have lower energy densities. As opposed to supply-on-demand, a renewable approach requires 'capture-when-available' and 'store-until-required' strategies. There are exceptions, such as hydroelectric schemes, but typically renewable systems function best at small scale near to point of use. They are not focused on a particular fuel type but use indigenous resources. Although a renewables approach is vulnerable to source intermittency, it does not have the same system vulnerabilities associated with the dominant model, for example top-down interconnected electrical systems are vulnerable to cascading faults, a regular occurrence in Europe and North America. Small-scale and distributed systems can be interconnected but the direction is typically horizontal, a structure not prone to cascading faults. Use of indigenous resources minimises geopolitical risks. Figure 1.8 shows the types of interventions needed to counter vulnerability and build resilience. In this sense, resilience is not the opposite of vulnerability but acts as an antidote.

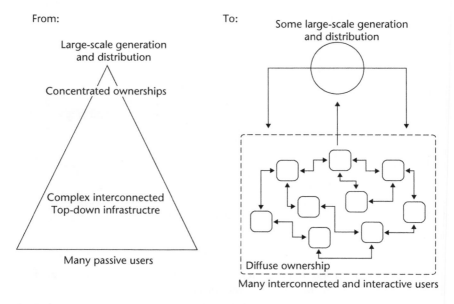

FIGURE 1.7 Conventional and distributed energy systems
Source: O'Brien and Hope (2010).

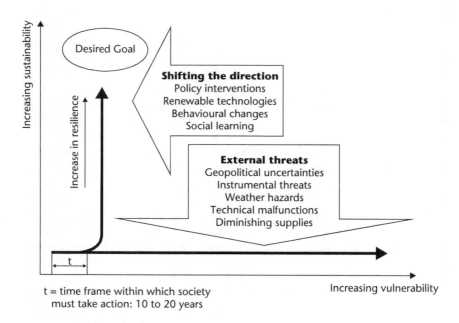

FIGURE 1.8 Shifting the direction
Source: O'Brien (2009).

There is evidence that some countries do have the instruments in place to increase resilience in energy systems, for example the EU; however, the EU has advocated a market-led, EU-wide interconnected system (EU Commission 2007). The United Kingdom has recently decided to develop new nuclear capacity. Both developments indicate that energy thinking is still locked by the status quo interpretation of sustainable development; however, the events in Japan at the Fukushima nuclear plant have made a number of countries review their nuclear plans.

What might happen?

In terms of what may happen in the future, there are plenty of doom merchants. Recent evidence of very peculiar weather conditions in many parts of the world suggests that the climate is changing, and IPCC, in its Fourth Assessment Report, attributes this directly to human interventions (IPCC 2007a). Despite the best efforts of well-funded climate change sceptic groups, unsurprisingly funded by corporations that have vested interests in fossil fuels, the reality we are facing means that the time for action should not be postponed (Hamilton 2010).

James Hansen, the director of the NASA Goddard Institute for Space Studies and a leading advocate of immediate climate action, argues in an editorial, worryingly titled *Game Over for the Climate*, that there is direct evidence for human impact on the climate that requires political action. He advocates a gradually rising carbon fee collected from fossil fuel companies, which would be wholly distributed to all Americans on a per-capita basis every month. He argues that this market-based approach would stimulate innovation, jobs and economic growth, and avoid enlarging government or having it pick winners or losers. Apart from the largest energy users, most Americans, he argues, would get more back than they paid in increased prices (Hansen 2012).

Though many support Hansen, his proposal plays into the hands of the ecological modernisers who place such faith in the market, but as the recent loss by JP Morgan Chase (an American multinational banking corporation) of some USD 2 billion in 2012 through very dubious trading deals shows, there is a problem that is deeply rooted in the market system. It can be argued that without effective regulation, a shift to a low carbon pathway will not be realised. Mechanisms such as CDM and JI (Joint Implementation) have not delivered carbon cuts, greater equity or social development. In fact, energy companies have gained huge windfall profits without cutting carbon emissions (House of Lords 2008). Cap and Trade, a system devised by those responsible for the sub-prime market, is preventing real progress in reducing carbon because many of these schemes allow offsetting against domestic targets (O'Keefe *et al.* 2010). In addition to the inherent problem in territorial accounting of global emissions between nation states, attempting to regulate emissions within a global context of unfettered markets essentially allows governments to 'offset' difficult decisions about *where* emissions cuts are made (While 2011; authors own emphasis). This is both political and fiduciary failure.

Renewable energy is all but unlimited, but transformation into usable energy supplies appears more costly than polluting fossil fuel alternatives. This hides the fact that the fossil fuel industry received USD 409 billion in handouts in 2010, compared with USD 66 billion for renewable technologies (IEA 2011a). The external costs of fossil fuel use are not reflected in price mechanisms. The market is distorted. This is evident in Cap and Trade, the mechanism lauded by many as a powerful tool to address climate change, but there is little evidence of its effectiveness. The EU Emissions Trading System (ETS) is claimed to have contributed 2 to 5 per cent carbon reductions in the pilot phase (2005–2007) (Ellerman and Buchner 2008; Ellerman et al. 2010). Others claim that any reductions were realised through carbon offsetting, meaning that no actual carbon reductions in the EU can be attributed to carbon trading (Gilbertson and Reyes 2009). The value of the traded volume in the EU ETS was USD 50 billion in 2007 and this rose to just under USD 120 billion in 2010 (World Bank 2011). This raises the question of effectiveness: would it not be more effective to use some of this money to invest in alternatives and adaptation?

Mol (2012) argues that the commodification of carbon can contribute to mitigation providing that state and non-state actors strongly advocate climate change mitigation. Given the track record of the financial sector and the political difficulties of 'banking' carbon credits, for example between Phase 2 and 3 of the EU ETS, it is difficult to imagine that carbon credits will not become toxic assets. It seems that efforts to stem the emission of GHGs have stalled and could fail. Development patterns throughout the world are being driven on the basis that there are no ecological limits, and we do not have an effective way of dealing with the problems that we will cause. There are a number of severe disconnections between policy frames that should be acting in harmony. Is it possible to find a way of framing policy that acts to ensure we do not destroy opportunities for this and future generations?

Can we change?

It would seem self-evident that we need to address the climate problem, but, as argued in this chapter, we seem to be mired in status quo thinking. A realistic programme for global mitigation seems unlikely and it is clear that there are other forces at work. We can accept that making international agreements, whether for climate change or any other global change issue, are extremely difficult. This is mainly due to the different and often conflicting agendas of the negotiators, who are themselves driven by a number of forces. There are three drivers that are also influencing those involved in climate change negotiations: demographics, lifestyle expectations and globalisation coupled with a misplaced faith in neo-liberalism.

The first issue is the growth in world population, which is expected to rise to some 9 billion by 2050, but the growth will be uneven with some countries experiencing hardly any growth while others will experience almost explosive growth. This is related to what is termed the demographic transition. In the United Kingdom prior to the Industrial Revolution, death and birth rates were roughly equal, meaning that the population, apart from immigration, remained fairly stable. As the Industrial Revolution advanced, developments in technology, particularly in food production and medicine, led to a fall in death rates – more babies survived into adulthood and people lived longer. It is clear that, at the time, people willingly accepted these new technological advances, but there was a considerable lag in the empowerment of women through education, cultural changes and access to contraception, which meant that birth rates remained high. By the time that both birth and death rates equalised, the overall population had increased. In the developed world today, predominantly the OECD countries, population levels are fairly stable; however, in the industrialising countries, such as India and China, this demographic transition is still continuing. The population expansion will be unevenly spread across the globe. Access to energy is vital to nations with growing populations. It is very unlikely that any political leader of a nation with a growing population will stand on a platform and campaign for current and future generations to have fewer energy resources in order to address climate change. This becomes even less probable when it is known that historically the developed world is primarily responsible for current GHG loadings and appears to be doing little to address the problem.

The second issue is related to the first. In today's world, the rapid expansion of global media has meant that the consumerist lifestyles of OECD nations have become the accepted norm of what a modern developed society should look like. Access to comfortable homes, good education and healthcare, leisure activities, welfare support, particularly for the elderly, access to consumer products and a variety of foodstuffs is seen as aspirational. There can be little argument that people in developing countries should not be denied the benefits that OECD societies enjoy – but this comes at an environmental cost. The garnering of resources, including energy, has led to extensive environmental degradation. This will continue. Already eyes are turning to the Arctic region in the hope that retreating ice sheets will allow access to new resources and transport corridors. The receding ice of Greenland is allowing resource exploration around its coastal areas and on newly exposed land. Is it likely that political leaders will restrain resource acquisition efforts for the sake of the environment?

The third issue is globalisation. Although this has been with us in one form or another ever since we learned how to trade, globalisation has become a powerful force post-Second World War that shapes how the world functions. Broadly, globalisation refers to the increasing international trade and capital flows, which also have political, cultural and ideological implications. The Bretton Woods conference in July 1944 was designed to address post-war reconstruction. The outcome of the conference was an agreement that nations should allocate 0.7 per cent of gross domestic product (GDP) to assist development in low-income countries; very few have actually honoured this agreement. It also established three institutions that have considerable influence today: the World Bank (initially IBRB – International Bank for Reconstruction and Development), the International Monetary Fund (IMF) and the World Trade Organisation (WTO initially GATT – General Agreement on Tariffs and Trade). These institutions championed post-war reconstruction and industrialisation in developing countries. They also promoted currency stabilisation by pegging them to the gold-standard. This ended in the 1970s when the United States removed the dollar from the gold-standard. This ended an era, termed controlled capitalism, which was replaced by neo-liberalism, during the Thatcher–Reagan era. Neo-liberalism championed the removal of tariffs and other controls on international trade and capital flows. This was reinforced by the Washington Consensus that advocated further reform through trade deregulation, foreign direct investment and privatisation of state enterprises. The implication of such reforms has been far-reaching. We have many global corporations that manufacture across the globe as they seek lower labour costs and/or access to cheap resources. They have been able to dictate to national governments the terms of conditions they want for their inward investments. Because of modern manufacturing methods, these companies can quickly set up new facilities in other countries should governments rail against the conditions imposed or if new opportunities become available. Similarly, banks and financial services institutions have globalised and are able to move money and financial instruments from bourse to bourse in a constant search for higher margins. Their only loyalty is to shareholders.

This brings us to our misplaced faith in neo-liberalism. The neo-liberalists claim that our interests are best served by maximising market freedom and minimising the role of the state. The free market, left to its own devices, will deliver efficiency, choice and prosperity. That has proven to be a chimera. From the sub-prime mortgage market to the massive problems of the Eurozone, the economic downturn in the OECD area has seen rises in unemployment, cuts to welfare services and, in some cases, the shrinking of the state. The fraudulent actions of the financial sector are now coming to light with the LIBOR (London Inter-Bank Rate) scandal, where major banks have manipulated this rate in order to make their institutions look healthier. Governments are fearful of taking action against the financial sector or of introducing financial regulations. Meanwhile, bodies such as the IMF prescribe more austerity. Neo-liberal thinking has failed to address the crisis facing many parts of the world. The promise of neo-liberalism is, at best, a hollow one. It has failed to deliver a better future. Unfettered free markets do not necessarily deliver the best solution. Their best solution is one that benefits shareholders. This is not always the best or most appropriate environmental solution. The actions needed to deliver climate solutions require intervention aimed at making the best environmental decision. This is unlikely to happen, as can be seen from the direction of the energy market.

Energy is just another commodity that is traded on the global market and, as long as there is demand, efforts both to acquire new sources and sell existing stocks will continue with little regard for the longevity of the resource. Although some analysts argue we have passed the peak oil point, where demand exceeds available or proved resources, there seems little sign of a slowdown in hydrocarbon use. A status quo argument would posit that we will find alternatives. There is some logic to this argument because there are sufficient renewable resources, but there has, in general, been little effort to develop these. The major energy conglomerates appear to have shifted their attention to other hydrocarbon resources. Although oil is projected to have a lifetime of 40 years and natural gas 70 years, other unconventional resources, such as shale and tar sands, are now seen as the next generation of hydrocarbons. The United States is projected to be self-sufficient in shale gas for the next 100 years and has considerable potential for shale oil. This will completely change the dynamics of international trade in oil and gas. With the prospect of energy supplies that are compatible with the current energy system, it appears that there is little incentive for climate negotiators to find a way of reducing fossil fuel use. Indeed, some would say that the lower carbon content of gas supplies means that there is a greater window of opportunity. Put simply, the negotiations are mired in status quo thinking. It appears the new thinking that needed to shift to renewables is missing internationally. Some efforts are being made, but the rise in population and expectations, driven by a globalising economy, will mean that we will continue to seek out the most cost-effective solutions, regardless of the environmental consequences.

This means that adaptation is now the only show in town. The question that remains, and it is a wicked one, is what level of temperature change we will have to adapt to and by when?

Our key argument is that the world is in disequilibrium and this drives change. We are now a major factor in the disequilibrium equation. Accelerated climate change is a major disruption to this process of change and that is our responsibility. Can we address this challenge by transforming the energy system and adapting our social, economic and political structures to ensure a more resilient and equitable future? It is impossible to make an absolute claim that we can. But it is human adaptability that has brought us to where we are now and we will have to use all of our adaptive capacity to address the problems we face.

In this book we attempt, first of all, to reject notions of resilience that emphasise a return to the status quo ante. In particular, we reject notions of resilience that say communities must stand by themselves. But, in building arguments for a resilience that allows opportunity to bounce forward, we have to explore the epistemologies that allow an understanding of change.

After a discussion of the history of climate change negotiations in the next chapter, we argue that we must place our understanding of climate science, particularly as it is used for policy formulation and negotiation, in a non-positivist tradition. Although we are no fans of the post-modernist discourse, we find the notion of a post-modernist science, i.e. one that cannot deliver answers to the questions, to be attractive. We seek to understand this discussion of the impact of climate change through an examination of what we understand to be the adaptation discourse. Here we are exploring adaptation under a capitalist mode of production. Under this mode of production, there are specific options to choose how we produce nature; however, this production of nature relates chiefly to issues of how we address poverty, and therefore development, at global and local levels. We briefly explore the dialectics that, in terms of people–environment relations, look backwards, remain anchored in the present, or look forward – and in looking forward, we briefly comment on substantive changes that are possible to decrease vulnerability by allowing equitable access to development. We end by outlining a schema that could allow a certain measurement of progress to a new world. Underlying all of this is a tension we all live and feel, namely what processes of democracy and governance encourage real gains in material life and an end to the vulnerability that underpins real risk in the modern world. Most institutions we mention seem not to be fit for purpose in addressing global need, sometimes it is an issue of scale, sometimes of purpose, sometimes of finance and sometimes of power. We do not have a universal answer to this problem but we hope clear messages emerge from this book.

The first message is do not allow resilience to become a language of oppression in which communities are essentially told by existing powers that they are on their own. Resilience building requires creating new opportunities of empowerment that allow people to address relationships between people and between people and nature. Secondly, in an imperfect world where the questions are more powerful than the evidence, let us think differently about the power of the questions, the purpose of the evidence, and most importantly, how political and academic debate can inform a widening circle of shared understanding of global risk.

2
THE CLIMATE JOURNEY

> A motorist lost in the Irish countryside sees a farmer leaning against a gate. He stops and asks, 'Is this the road to Dublin?' The farmer replies, 'If I were you, I wouldn't start from here.'
>
> (Anecdote)

Introduction

Like the motorist in the Irish countryside, the climate journey appears, at times, to have lost its way. Adaptation did not get off to a good start and, as the farmer implies, appears to be more of an afterthought. Although it was part of the journey, it was very much the backseat passenger. In retrospect, this seems odd because adaptation to various stimuli has been an integral part of human evolution; however, the early emphasis of climate talks focused on adaptation of the energy system to reduce GHG emissions as a mitigation strategy to reduce the impacts of climate change (Schipper 2006). Mitigation was in the driving seat. Little thought, it seems, was given to the possible consequences of emissions already present in the atmosphere (Bowerman et al. 2011). When the UNFCCC was established, this skewed emphasis was embedded. The skewing of the Climate Convention has prevented progress on adaptation and it is only recently that there has been recognition of the need to adapt to historic emissions and acknowledgement that we will have to adapt to ongoing emissions. The failure of the international effort to formulate a successor to the Kyoto Protocol, despite all of its shortcomings, is driving us into an unknown set of climate conditions because there is little indication to suggest that we can agree to stabilise GHG concentrations in the atmosphere (Bowerman et al. 2011; Anderson and Bows 2011). It could be a very bumpy future.

If anything can be learned from the climate journey it is this: trying to get an international agreement with real commitments and real teeth is like herding cats. Negotiations have been protracted and at times hostile. Some actors have thrown their toys out of the pram or simply taken them home. After a protracted effort, the Kyoto Protocol was agreed and ratified. That says a lot about the tenacity of the negotiating teams; however, even its modest targets have proved too difficult for some signatories. Given the growing amount of evidence that supports the view that human activities are adversely affecting the planetary climate system, it seems perverse that the international community cannot find a way of agreeing a target and a timetable to reduce GHG emissions – but that was precisely the outcome of the Durban meeting in 2011. The only tool we have left to deal with climate-driven changes is adaptation. The question now becomes how did we get to this juncture and can we adapt quickly enough and with purpose to counter future changes?

Where is adaptation?

Adapting to the consequences of climate change, though recognised by the Convention, negotiations lacked the focus of the mitigation debate. In the early negotiations this was, at least partially, due to a widely held perception that focusing overly on 'adaptation' was a defeatist position because the priority had to be on reducing GHGs (Schipper 2006). Furthermore, Article 4.8 of the Convention, the basis for negotiations on adaptation, refers both to the needs and concerns of developing countries vulnerable to climate change and to the adverse effects of climate protection measures on oil exporting countries. This perverse link between those countries vulnerable to the effects of climate change and those vulnerable to the impact of reducing reliance on fossil fuels as a climate response effectively prevented meaningful progress on adaptation in the early phase of the Convention.

The Buenos Aires Plan of Action (BAPA) of 1998, agreed at COP 4, established a package of actions to minimise impacts of climate change. This decision effectively mainstreamed adaptation into the UNFCCC process (Dessai and Schipper 2003). It was important for two reasons. First, it established funding mechanisms for adaptation (Table 2.1). Second, it marked the start of a strategy for adaptation work. Adaptation efforts were focused on countries in geographic locations most vulnerable to the impacts of climate change, such as drought-prone areas or low-lying areas vulnerable to flooding, and Least Developed Countries (LDCs) because they have lower adaptive capacities than developed countries because of limited financial resources, skills and technologies, high levels of poverty and a high reliance on climate-sensitive sectors such as agriculture and fishing.

The funds established by COP 4 were:

- **Special Climate Change Fund:** for adaptation activities, programmes and measures in developing countries, planning preparedness and disaster relief, rapid response to extreme weather, but also for technology transfer and activities to help diversify economies of oil-producing countries (UNFCCC 2002a).
- **Least Developed Countries Fund:** for assessing adaptation needs through National Adaptation Programmes of Action (NAPAs) (UNFCCC 2002a).
- **Adaptation Fund:** funding for adaptation projects for countries that are parties to the Kyoto Protocol (UNFCCC 2002b).

Table 2.1 summarises the funds that have been made available under UNFCCC.

TABLE 2.1 Summary of adaptation funding under UNFCCC

Fund	Created under	Does the project have to demonstrate global environmental benefits?	Beneficiaries	Funding sources	COP and GEF
GEF Trust Fund	UNFCCC	Incremental cost to achieve global environmental benefits	Developing countries	GEF	1/CP.1 5/CP.7 GEF/C.23/Inf.8
GEF Strategic Priority on Adaptation (small grants fund)	UNFCCC	Incremental cost to achieve global environmental benefits	Developing countries	GEF	6/CP.7 GEF/C.23/Inf.8
Special Climate Change Fund	UNFCCC	Additional cost of adaptation measures; uses a sliding scale	Developing countries	Developed countries discretionary pledges	5/CP.7 7/CP.7 5/CP.9 GEF/C.24/12 GEF/C.25/4/Rev.1
Least Developed Countries Fund	UNFCCC	Additional cost of adaptation measures; uses a sliding scale	Least developed countries	Developed countries discretionary pledges	5/CP.7 7/CP.7 27/CP.7 28/CP.7 29/CP.7 6/CP.9 GEF/C.24/Inf.7 GEF/C/25/4/rev.1 Decision 11/CP.11
Adaptation Fund	Kyoto protocol	No	Developing countries	Share of proceeds from CDM; other sources	5/CP.7 10/CP.7 17/CP.7 3/CMP.1 28/CMP.1 5/CMP.2 1/CMP.3

Source: Mace (2005: 231); Grasso (2010: 83).

Funding for adaptation under UNFCCC and the Kyoto Protocol amounted to some USD 310 million up to 2007 (Reid and Huq 2007). The Strategic Priority on Adaptation, with USD 50 million of Global Environment Facility (GEF) funding, supports demonstration projects concerned with the management of ecosystems to show how climate change adaptation planning and assessment can be practically integrated into national policy and sustainable development planning. The Least Developed Country Fund supports the development of NAPAs and implementation of their NAPA projects. The Special Climate Change Fund is for adaptation activities in all developing countries. In addition to these funds, donors have provided bilateral funding of around USD 50 million for adaptation activities for over 50 adaptation projects in 29 countries.

The Special Climate Change Fund and the Least Developed Countries Fund are supported by donor contributions. The Adaptation Fund is, controversially, supported by a 2 per cent levy on projects generating emission credits through the CDM. The controversial aspect of the CDM levy for adaptation arises essentially because this is a 'tax' on developing country investments whereas JI and the EU ETS (developed world investments) are not subject to a similar levy, as well as the fact that each CDM project increases the emissions from developing countries (Grasso 2010; Mace 2005). The responsibility for the administration of the funds was given to the GEF, an independent financial organisation established in 1991 to help developing countries fund projects and programmes that protect the global environment. GEF grants support projects related to biodiversity, climate change, international waters, land degradation, the ozone layer and persistent organic pollutants.

Marrakesh represented a watershed in the evolution of the adaptation and climate debate since the inception of UNFCCC. The Marrakesh Accords, agreed at COP 7 in 2001, are significant in terms of mitigation and adaptation. The Accords established the 'rule-book' for the Kyoto Protocol, effectively paving the way for UNFCCC members to become signatories to a legally binding instrument explicitly targeted at reducing global GHG emissions. They also dealt with the issue of 'sinks' and established the rules for emissions trading under the Protocol and for project-based approaches under the CDM and JI. One issue that remained unresolved at Marrakesh was the differentiation between countries that are adversely impacted by a changing climate and those that would be economically impacted by response measures.

The Kyoto Protocol did not come into force until 2005 when it was ratified by Russia. From that point, the COP also hosted the Meeting of the Parties to the Kyoto Protocol (MOP); however, the effectiveness of the Protocol was severely undermined because the US Senate, then representing the world's largest GHG emitter, the United States, had refused to ratify unless there were similar commitments from developing countries. Some claimed the Protocol's target of around 5 per cent reduction was too modest. Others argued that it was important because it would determine if the Cap and Trade approach of the Protocol could serve as viable model for future agreements post-Kyoto (Ott et al. 2005).

COP 8 in 2002 and the Delhi Ministerial Declaration on Climate Change and Sustainable Development made adaptation a high priority for all countries, but particularly for LDCs and Small Island Developing States (SIDS). Arguably, this declaration gave adaptation equal status with mitigation, although some argue that adaptation was used as a mechanism to deflect attention away from developing country mitigation commitments (Schipper 2006).

At COP 9 in Milan, the Special Climate Change Fund (SCCF) was operationalised with an 'adopted decision'[1] that gave top priority to adaptation activities and emphasised that adaptation should be country-driven, cost effective and integrated into national sustainable development and poverty-reduction strategies. National Communications or LDC NAPAs were agreed as the vehicle for informing on implementation. This was an important step as anthropogenic-driven climate changes were being detected at the regional and ecosystem scale (Karoly et al. 2003; Parmesan and Yohe 2003; Root et al. 2003). No agreement, however, was reached on the Adaptation Fund or LDC Fund at COP 9 because of ongoing differences between the genuine adaptation concerns for LDCs and concerns held by oil producing countries of the impact of response measures.

This was finally resolved in 2004 at COP 10, Buenos Aires, in the Buenos Aires Programme of Work on Adaptation and Response Measures. The agreement specified the means for implementing the decision on adaptation measures reached at COP 7 in Marrakesh. On adaptation to climate impacts, the Buenos Aires Programme spelled out activities to improve data collection, strengthen training and in-country capacity, undertake pilot projects and promote technology transfer, and called for three regional workshops and an expert meeting during the following 2 years to help identify adaptation needs and concerns. For adaptation to response measures, the decision calls for expert meetings on ways to promote economic diversification in oil-producing countries. The decision also requested the Subsidiary Body for Scientific and Technological Advice (SBSTA) to develop a 5-year work programme to further advance work on adaptation to climate impacts. Effectively, COP 10 (sometimes referred to as the Adaptation COP) separated adaptation under the Convention (adverse effects) from the impact of response measures on oil producing countries (Ott et al. 2005).

The 5-year programme developed by SBSTA was agreed at COP 11 in Montreal, 2005. The programme aimed to assist parties to improve their understanding of adaptation, impacts and vulnerability, and to make informed decisions on practical actions and measures. To help parties better assess their vulnerability, the programme promoted improved vulnerability assessment tools, climate monitoring and projections, and understanding of variability and extreme events. To support adaptation planning and action, the programme promoted analysis and sharing of adaptation measures, research on adaptation technologies and development of economic diversification strategies. The work has been conducted primarily through workshops, expert groups and technical papers.

COP 12, held in 2006 in Nairobi, made some modest progress on adaptation. The 5-year work plan agreed in Buenos Aires was renamed the Nairobi Work

Programme on Impacts, Vulnerability and Adaptation to Climate Change. The Programme calls for a series of workshops and reports over the following 3 years to share and analyse information on topics, including climate data and modelling, adaptation tools and methods, climate variability and extreme events, and economic diversification.

The issue of the administration of the Adaptation Fund had been raised a number of times since Marrakesh (Mace 2005; Grasso 2010). As the fund is supported by a levy on projects generating emission credits, many developing countries argue that it should be managed not by the donor-dominated Global Environmental Facility, which manages the other funds, but by an entity giving developing countries more say in decision making. No agreement was reached on the choice of entity; however, an agreement on governance principles was reached. Funding would be made available for national, regional and community level efforts, and that whatever governing body is selected, the majority of its members will represent developing countries.

Road map for post-Kyoto negotiations

The G8 Summit in Heiligendamm, June 2007, was the scene of calls from world leaders requesting the UNFCCC to agree on a road map by 2009 for the negotiations to succeed the Kyoto Protocol, which expired in 2012. Momentum was gained as further calls joined from both the High Level Event on Climate Change and the Major Economic Meeting on Climate Change and Energy Security. The desire expressed amongst nations to reach an agreement that would ensure continuity in the current climate change regime led to high expectations for a breakthrough at the UNFCCC Conference (COP 13, MOP 3) in Bali, December 2007.

The main objectives of the Bali Conference were to launch the 'Bali Road Map' (BRM) (UNFCCC 2007a), reach agreement on the proposed Adaptation Fund, discuss ways by which to reduce emissions from deforestation, discuss carbon market issues and discuss arrangements for review of the Kyoto Protocol.

The BRM is not one single document but a collection of 25 decisions and one resolution highlighting the agreement reached on a number of issues that would underpin the negotiations over the next 2 years in the run-up to Copenhagen via Poznan. The Bali Action Plan (BAP), a component of the BRM, is a four-page document that highlighted the four pathways that the negotiations should follow: mitigation, adaptation, technological cooperation and financial support. These pathways are part of a comprehensive process to enable the full, effective and sustained implementation of the Convention through long-term cooperative action, up to and beyond 2012, in order to reach an agreed outcome and adopt a decision at the COP 15 (Copenhagen 2009). These pathways are supposed to form the four building blocks of the post-2012 regime, and they were split into a two-track process to enable the United States to remain a part of the post-2012 negotiations. Therefore, the Ad Hoc Working Group on Further Commitments for Annex I Parties under the Kyoto Protocol (AWG-KP) was established, together with the Ad Hoc Working Group on Long-term Cooperative Action under the Convention (AWG-LCA), which included the United States and other major emitters who are not party to Kyoto. These Ad Hoc Working Groups, created under the BAP, were charged with conducting this process and presenting the progress at the COP 15, arguably deflecting priority away from the Kyoto Protocol (Grubb 2011).

Arguably, the BAP and the BRM give climate change adaptation the same importance as mitigation. Decision 1 calls for 'enhanced action on adaptation', including international cooperation in the implementation of adaptation actions that take into account the vulnerable position of LDC, SIDS and some African states, risk management and reduction strategies, disaster reduction strategies and economic diversification.

Decision 1 under the BRM also included the launch of the Adaptation Fund. Under this decision, developing countries' parties to the Kyoto Protocol, which are particularly vulnerable to the adverse effects of climate change, will have access to funding for adaptation. The purpose of the Adaptation Fund was to finance concrete adaptation projects and programmes that are country driven and based on the needs, views and priorities of eligible parties. The Fund was administered initially by the GEF, with the World Bank acting as trustee. A 16-member Adaptation Fund Board (AFB) would perform the supervisory and management functions. The members were selected from both industrialised and developing countries and are accountable to the COP.

This left less than 2 years from the adoption of the BRM in December 2007 until the COP 15 in Copenhagen in 2009 to reach a consensus on a new post-Kyoto agreement. That was a short amount of time for international negotiators to draft legal text that could be agreed by different nations who have radically different positions in the climate change negotiations. The first step taken was the Bangkok Climate Change Talks in March 2008. The objective of these talks was to develop a working programme around which to structure the 2-year negotiations. The programme was drafted by the AWG-LCA and each of the four pathways was discussed. During the Bangkok Talks, it was decided that initial work would consider how to advance action on adaptation through technology and finance. All in all, the first milestone on the way to Copenhagen was reached.

The second step followed with the Bonn Climate Change Talks in June 2008. This event involved meetings of both the SBSTA and SBI (Subsidiary Body on Implementation), and also the second session of the AWG-LCA. These talks aimed at finding ways to advance adaptation through finance and technology, enabling informed debate on these issues and gaining a clearer understanding among delegates of what countries would ultimately like to see contained within a long-term agreement. Based on the discussions, the AWG-LCA put forward several proposals on what parties envisaged a final agreement containing. This was seen as a positive step, but it was felt that the proposals needed to take a more specific textual form if they were to provide the basis of a negotiated text in COP 14 that could then be adopted at COP 15. Some progress was also made on the adaptation front, in that parties agreed to implement the second phase of the Nairobi Work Programme on impacts, vulnerability and adaptation to climate change.

The Accra Climate Change talks held in August 2008 again attempted to take forward work on reaching a post-2012 agreement. The AWG-LCA met for a third time, as did the AWG-KP, which looked at emission rules and tools under the Protocol.

The COP 14 in Poznań (December 2008) was the final meeting before a new agreement could be adopted at the COP 15 in 2009. The major issues addressed at Poznań included technology transfer, finance and practical adaptation actions. Mitigation was considered through a review of the implementation of the Kyoto Protocol and through the discussion of the possible actions to be taken by developing countries towards satisfying their post-2012 obligations. This meeting also served as an important opportunity for an assessment of progress made throughout the year. The conference concluded with a clear commitment from governments to shift into full negotiating mode the following year in order to shape an ambitious and effective international response to climate change to be agreed at the COP 15 in Copenhagen at the end of 2009. Parties agreed that the first draft of a concrete negotiating text would be available at a UNFCCC gathering in Bonn in June 2009.

The Copenhagen Climate Change Conference was the fifteenth meeting of the parties under the UNFCCC since the first in 1995. It was intended to act as the defining moment in whether or not the international community could achieve a smooth and immediate transition from the regime under the Kyoto Protocol to one governed by an updated 'Copenhagen Protocol'. The main goal for the COP 15 was to adopt a global climate change agreement under the auspices of the UNFCCC that adequately addressed both updated scientific knowledge and the four pathways as outlined in the BAP; however, Copenhagen was destined to end in disarray.

Copenhagen to Cancún

It is clear that in the run-up to Copenhagen expectations were high. Groups such as the World Wildlife Fund (WWF) set out their expectations in March 2009 (WWF 2009). But reports in February 2009 indicated trouble when the Chinese Premier, Wen Jiabao, said that it would be difficult for the Chinese to accept quantified emission reduction quotas because the country was still in the early stages of development. Indian officials had already indicated that they were unlikely to agree to a mandatory cap on emissions in Copenhagen, while Russia's Prime Minister, Vladimir Putin, had taken a similarly non-committal stance on climate change in his keynote address at the Davos forum (Young 2009). Surprisingly, in November 2009, China announced that it would reduce the intensity of GHG emissions over the next decade, saying that by 2020 it would reduce its carbon dioxide emissions per unit of GDP by 40 to 45 per cent compared with levels in 2005. This rekindled hope that perhaps Copenhagen could deliver (Pierson and Tankersley 2009). The White House announced in November that President Obama would attend the

Copenhagen meeting. This stirred further hopes that the United States might be able to give the talks further momentum because the United States indicated that it would commit to cutting emissions by 17 per cent by 2020 (Sheppard 2009). Despite surges in optimism, it was becoming clear that there was little prospect of an agreement, with experts in the United States suggesting the best that could be achieved was to develop some building blocks for future negotiations (Miller 2009). To compound the problems in the run-up to Copenhagen, a leak of embarrassing, and in some cases troubling, emails from a major global climate centre, East Anglia's Climatic Research Unit (CRU), was orchestrated by sceptics of human-caused global warming (Her Majesty's Government 2010). This put even more pressure on leaders to get a deal (or at least the beginnings of one) in Copenhagen. The Director of the unit, Phil Jones, who was at the centre of the storm, did stand down temporarily from his post. He was later exonerated of blame.

There is little doubt that the 'ClimateGate' incident strengthened the hand of the climate deniers, but it is not clear if this had any impact on the Copenhagen meeting. There has been a decline in support for climate change both before and after Copenhagen, but this could easily be associated with the conflicting signals in the run-up to Copenhagen, the global financial crisis and the disappointing outcome and lack of leadership shown by global leaders (Adam 2010). This is somewhat surprising because the IPCC had clearly supported the view that human activity was impacting the climate in its Fourth Assessment Report (IPCC 2007a).

In the run-up to Copenhagen, a group of authors, primarily previous IPCC lead authors familiar with the rigor and completeness required for a scientific assessment of this nature, produced *The Copenhagen Diagnosis* (The Copenhagen Diagnosis 2009). The purpose of this report was two-fold. First, it acted as an interim evaluation of the evolving science midway through an IPCC cycle (IPCC AR5 was not due for completion until 2013). Second, the report served as a handbook of science updates that supplemented the previous IPCC report. Despite some alarming findings (see Box 2.1), the report seems to have had little impact.

BOX 2.1

SUMMARY OF FINDINGS OF THE COPENHAGEN DIAGNOSIS

Surging GHG emissions. Global carbon dioxide emissions from fossil fuels in 2008 were 40% higher than those in 1990. Even if global emission rates are stabilised at present-day levels, just 20 more years of emissions would give a 25% probability that warming will exceed 2 degrees Celsius, even with zero emissions after 2030. Every year of delayed action increases the chances of warming exceeding 2 degrees Celsius.

Recent global temperatures demonstrate human-based warming. Over the past 25 years temperatures have increased at a rate of 0.19 degrees Celsius per decade in every good agreement with predictions based on GHG increases. Even over the past 10 years, despite a decrease in solar forcing, the trend continues to be one of warming. Natural, short-term fluctuations are occurring as usual, but there have been no significant changes in the underlying warming trend.

Acceleration of melting of ice sheets, glaciers and ice caps. A wide array of satellite and ice measurements now demonstrates beyond doubt that both the Greenland and Antarctic ice sheets are losing mass at an increasing rate. Melting of glaciers and ice caps in other parts of the world has also accelerated since 1990.

Rapid Arctic sea-ice decline. Summer melting of Arctic sea ice has accelerated far beyond the expectations of climate models. The area of summer sea ice during 2007–2009 was about 40% less than the average prediction from IPCC AR4 climate models.

Current sea level rise underestimated. Satellites show the great global average sea level rise (3.4 millimetres/year over the past 15 years) to be 80% above past IPCC predictions. This acceleration in the rise of sea level is consistent with a doubling in contribution from melting of glaciers, ice caps and the Greenland and West Antarctic ice sheets.

Sea level prediction revised. By 2100, global sea level is likely to rise at least twice as much as projected by Working Group 1 of the IPCC AR4; for unmitigated emissions it may well exceed 1 metre. The upper limit has been estimated at a 2-metre sea level rise by 2100. Sea level will continue to rise for centuries after global temperatures have stabilised and several metres of sea level rise must be expected over the next few centuries.

Delay in action risks irreversible damage. Several vulnerable elements in the climate system (e.g. continental ice sheets, Amazon rainforest, West African monsoon and others) could be pushed towards abrupt or irreversible change if warming continues in a business-as-usual way throughout this century. The risk of transgressing critical thresholds ('tipping points') increases strongly with ongoing climate change. Thus waiting for higher levels of scientific certainty could mean that some tipping points will be crossed before they are recognised. The turning point must come soon: if global warming is to be limited to a maximum increase of 2 degrees Celsius above preindustrial values, global emissions need to peak between 2015 and 2020 and then decline rapidly. To stabilise climate, a decarbonised global society – with near-zero emissions of CO_2 and other long-lived GHGs – needs to be reached well within this century. More specifically, the average annual per capita emissions will have to shrink to well under 1 metric tonne CO_2 by 2050. This is 80–95% below the per capita emissions in developed nations in 2000.

Source: The Copenhagen Diagnosis (2009).

So what did happen in the Copenhagen meeting? Connie Hedegaard, the conference president, resigned on 16 December 2009. She claimed that this was just a procedural matter because a large number of heads of state and government were arriving and it would be more appropriate for the Danish Prime Minister, Lars Løkke Rasmussen, to preside over the final stages of the talks (Stratton and Vidal 2009). Rumours, however, abounded that the real reason for her resignation was the so-called Danish text. This had been drafted by Prime Minister Rasmussen and presented to a few countries (the circle of commitment that included the United Kingdom, the United States and Denmark) a few weeks before the meeting officially started. The idea was to present the paper in the event of a deadlock, which was thought to be very likely. The document was leaked to *The Guardian* newspaper. The perception was that the text clearly favoured the United States and the West, and was also interpreted by developing countries as setting unequal limits on per capita carbon emissions for developed and developing countries in 2050; in short, rich people would be entitled to emit twice as much as the poor (Vidal 2009).

An analysis of the text by developing countries shows why there was such deep unease over details of the text. In particular, the text intended to:

- force developing countries to agree to specific emission cuts and measures that were not part of the original UN agreement;
- divide poor countries further by creating a new category of developing countries called 'the most vulnerable';
- weaken the UN's role in handling climate finance; and
- not allow poor countries to emit more than 1.44 tonnes of carbon per person by 2050, while allowing rich countries to emit 2.67 tonnes.

Developing countries that saw the text were furious that these conditions were being promoted by rich countries without their knowledge and without discussion in the negotiations. This is a clear example of the developed world being mired in status quo thinking.

The events at the meeting have to be contextualised against the background of other rumours and allegations, for example there were allegations that the United States and China had been in backdoor discussions since 1997 on forging an agreement prior to the Copenhagen meeting (Goldenberg 2009). Rumours spread from diplomatic cables, leaked by WikiLeaks, which suggested that the United States had been trying to manipulate others into supporting the Copenhagen Accord (Carrington 2010a). It is not possible to verify the impact this had on the post-meeting period and the run-up to Cancún, but despite all of the negativity surrounding the process from Kyoto to Copenhagen, it is worth reflecting on whether anything was achieved.

Wonderful, Wonderful Copenhagen

The above opening line of the song 'Wonderful Copenhagen' by Frank Loesser from the 1952 film *Hans Christian Andersen* is probably not what many who attended the COP 15 felt. The expectations for some sort of agreement to succeed the Kyoto Protocol were dashed, but progress was made in some areas despite failures in others.

The Copenhagen Accord did not legally commit any countries to cut emissions, it simply noted the need for deep cuts to keep the average global temperature increase below 2 degrees Celsius – despite a general understanding that this is not the upper limit to avoid 'dangerous climate change' (Anderson and Bows 2011: 20). It acknowledged the need for funding to support what it described as the commitment by developed countries to provide new and additional resources, including forestry and investments through international institutions, approaching USD 30 billion for the period 2010–2012, with balanced allocation between adaptation and mitigation. In the longer term, it committed developed countries to a goal of jointly mobilising USD 100 billion dollars a year by 2020 to address the needs of developing countries (The Copenhagen Accord 2009). At least this was recognition of a shared problem (Hassing et al. 2009; Parry et al. 2009), but little clarity was given on how these funds would be raised from international sources based on the OECD recognised 'polluter pays' principle and, more importantly, how they would be channelled (Atteridge 2011).

Another success was the issue around land-use forests. The principles for a framework for Reducing Emissions from Deforestation and Forest Degradation (REDD) were established, reigning in large timber ranching agriculture and mining interests and providing some protection for the rights of indigenous people. Again, however, much work has to be done on quantitative issues and compensation mechanisms.

There are some lessons that can be drawn from the Copenhagen experience. For much of the last 20 years there has been talk of a uni-polar world, led by the United States, pushing an agenda of free market globalisation under what is frequently described as the Washington Consensus. The emphasis has been on Washington's leadership rather than the fact that it was only a consensus in Washington. In fact, what has been emerging strongly for some time is the emergence of a multipolar world. Obviously, there are the giants of the twenty-first century – the Asian century, namely China and India, together with Japan and Korea. There are the voices of the dispossessed of Africa, led by South Africa, Nigeria and increasingly Ghana; and then there are Brazil and Argentina leading the Latin and Central American position. In the North, there are a distinct series of blocks between the United States and Europe, and within a broader Europe there is the negative Russian influence on the Euro–Asian periphery. It is within these blocks that progress has to be found.

Within the different groupings of the UNFCCC, there are three key groupings (nation states, businesses and civic society) that operate to radically different rules. Countries defend the nation state within national boundaries and work to influence others. Businesses do not recognise borders aside from tax purposes and international trading regimes, such as those driven by the WTO, and they work to meet the requirements of shareholding capital. Civic society, once defined via the nation state, is more globalised as a result of global environmental issues and rapid technological developments in media, communications and travel. Often different communities may hold common values within and across frontiers. In general, civic society can be characterised by the desire to promote well-being, whether locally or globally focused. There is no definitive structure or pattern of organisation for civic society's rather disparate voices; however, when united in a common cause, civic society can be a powerful agent of change and has a demonstrated ability to contribute to global governance.

Onwards from Copenhagen: to Cancún and Durban

In the run-up to Cancún, it is no surprise that expectations for the Cancún meeting were low given the outcome of Copenhagen. There were four preparatory rounds of negotiations held during the run-up to Cancún. The first three were held in Bonn in Germany from 9 to 11 April, 31 May to 10 June (in conjunction with the 32nd sessions of SBSTA and SBI) and from 2 to 6 August 2010. By the time the negotiations opened, some 110 countries had backed the Copenhagen Accord.

The aim of the negotiators who met in April was to pick up the pieces from Copenhagen to see what could be salvaged and turned into a proper global agreement for the Cancún meeting. The talks were mainly procedural, for example which texts to start with, how many meetings to hold before Cancún, whether to mandate the chair to prepare draft texts, and so on. Although this first preparatory meeting was mainly procedural, in the final plenary the outgoing head of the UNFCCC, Yvo de Boer, said there was no chance of reaching a final deal by November in Cancún. He claimed that the United States did not want a second round of Kyoto and that it was likely that it may need two separate legal treaties to bring together the United States and developing countries. The meeting ended with a division between those who backed the United States in wanting the Copenhagen Accord and other countries, led by India and China, recalling the necessity to follow the texts already under discussion in the UN framework (Huq 2010; International Council for Local Environmental Initiatives (ICLEI) 2010; Vidal 2010).

The June meeting in Bonn comprised a number of meetings of the subsidiary bodies of UNFCCC and the AWG-KP and AWG-LCA. One of the key issues under the subsidiary bodies was an agenda item under the SBSTA on scientific, technical and socio-economic aspects of mitigating climate change. The Alliance of Small Island States (AOSIS), with most other parties, requested a technical paper on options for limiting global average temperature increase to 1.5 degrees Celsius and 2 degrees Celsius from pre-industrial levels – an issue pressed at Copenhagen. This was opposed by Saudi Arabia, Oman, Kuwait and Qatar. Eventually no agreement was reached and the original agenda was adopted. Many expressed their deep disappointment. Many developing countries believed that a 2-degrees Celsius increase would adversely impact poorer nations. Despite this disappointment, there was a general feeling that some progress was being made. At the end of the meeting delegates bid farewell to the outgoing UNFCCC Executive Secretary Yvo de Boer and welcomed the appointment of Christiana Figueres from Costa Rica as the next Executive Secretary (IISD RS 2010).

The third preparatory meeting in Bonn focused on mitigation for both developed and developing countries as well as institutional issues for financing. The outcome of the meeting was not positive, with the Executive Secretary Christiana Figueres stating in a press release that many countries had reinserted established positions into the texts, increasing the number of options for action, 'To achieve desired outcomes in Cancún, governments must radically narrow down the choices on the table.' (UNFCCC 2010: 2) This could tie up the meeting in Cancún; however, the US Chief Negotiator, John Pershing, claimed that some countries were back-tracking over agreements made in Copenhagen. The 43-nation AOSIS claimed that the lack of legislation in the United States to curb emissions was a setback and that the rich nations' pledge for cuts was not enough. Connie Hedegaard, the EU's Climate Commissioner, was pessimistic claiming that, if anything, the talks had gone backwards (BBC 2010).

The fourth and final preparatory meeting was held in Tianjin in China from 4 to 9 October 2010. Some progress was made on the issues of forestry, technology transfer and financing for adaptation for poor countries, but the meeting was overshadowed by a clash between the United States and China. The United States accused developing countries of not building on the Copenhagen Accord and China accused the United States of not providing finance or technology but insisting that poorer countries accept stringent monitoring and voluntary domestic actions. In short, the United States was doing very little but expected others to deliver what it could not – or would not – deliver. There were some hopes that the meeting in Cancún could agree details of a fund to transfer USD 100 billion (£63 billion) a year from rich countries to help poor nations cope with the projected consequences of climate change, but developing nations described the sum as substantial but inadequate (Harrabin 2010; Watts 2010).

This can hardly be described as a positive run-up to the Cancún meeting, which took place from 29 November to 10 December 2010. One of the many challenges for the Cancún meeting was to make progress on bringing the building blocks of the Copenhagen Accord into an acceptable package, such as detailed agreements on climate finance, preventing deforestation, enabling technology transfer and accounting for emission cuts by emerging economies such as China and India. Even those modest ambitions were put in jeopardy when Japan and then Russia announced they would not sign up to a second term of the Kyoto Protocol unless the world's big emitters, China and the United States, were also legally bound to action. There was concern that the UN process could unravel, effectively leaving no vehicle in which international agreements could be made; however, unlike Copenhagen, the talks did not descend into acrimony and dissent, largely due to the effective diplomacy of Patricia Espinosa, the Mexican Foreign Minister, who presided over the talks (Carrington 2010b).

So did the Cancún meeting actually achieve anything? There is general, if somewhat grudging acknowledgment, that some progress was made with, for example non-governmental organisations (NGOs) such as Greenpeace, Friends of the Earth and WWF-UK claiming that although the agreement was weak it was probably the best that could be achieved given the circumstances. But that did not disguise the fact that much more was needed. Representatives from country after country acknowledged the agreement was not perfect, but they supported it as progress towards a final deal. Bolivia though hit out at the proposals, likening them to genocide (Beament 2010). It is worth reflecting in more detail on what was achieved at Cancún.

Emissions

The Cancún Agreements made no changes to the magnitude of the voluntary emissions reduction commitments made under the Copenhagen Accord, either for developed or developing countries. But the Cancún Agreements did urge the developed countries to increase the ambition of their targets to a level that was consistent with the latest recommendations of the IPCC. Furthermore, developed countries were urged to prepare low-carbon development strategies or plans. The Agreements encouraged developing countries to do so as well, though there is no process established to define them (Decision1/CP.16). The Cancún Agreements extended negotiations and this ensured that there was no gap between the first and second commitment periods under UNFCCC until the next COP in Durban, South Africa. Effectively the Cancún Agreements kept hope alive that the UNFCCC goal of legally binding emissions reduction targets can be achieved in future negotiations. The Agreements also kept the Copenhagen target of aiming for less than a 2-degree Celsius rise in average overall global temperature. It also agreed to investigate strengthening this target to 1.5 degrees Celsius.

Under the Cancún Agreements, developing countries agreed to take Nationally Appropriate Mitigation Actions (NAMAs), supported by technology and finance, that are aimed at achieving a deviation in emissions relative to 'business as usual' emissions in 2020. The Agreements also called for workshops to clarify the assumptions behind countries' mitigation pledges and, in the case of developed countries, to consider ways to increase their level of ambition. They also established a two-part registry. In the first part, which is intended to facilitate matching of developing country actions with support, developing countries can list proposed actions in need of support, and developed countries can list support available or provided. The second part will record all developing country NAMAs – whether supported or unsupported.

Measurement, reporting and verification under the UNFCCC

To strengthen the measurement, reporting and verification (MRV) of mitigation actions and support for developing countries, the Agreements call for:

- more detailed reporting on mitigation actions and support provided or received in the national communications of both developed and developing countries;

- guidelines for international MRV of mitigation actions receiving international support, and 'general' guidelines for domestic MRV of autonomous actions for developing countries; and
- new biennial reports by developed countries on emission reduction progress and support provided, and for developing countries GHG inventories, mitigation actions, needs and support received.

The Cancún Agreements established new processes within the SBI to consider mitigation efforts – called 'international assessments' for developed countries, and 'international consultations and analysis' for developing countries. For the latter, the decision specifies that the process should:

- be 'non-intrusive, non-punitive and respectful of national sovereignty focus' on unsupported actions;
- not consider the 'appropriateness' of a country's domestic policies; and
- include an analysis by technical experts with results set out in a summary report.

Finance for developing countries for their obligations under the UNFCCC

The Cancún Agreements have incorporated the finance goals set out in the Copenhagen Accord (developed countries' commitment to provide USD 30 billion in fast-start finance for developing countries in 2010–2012 and to mobilise USD 100 billion a year in public and private finance by 2020 for adaptation and low carbon initiatives); however, there is no indication of where this funding will come from, apart from non-binding pledges made by some countries. The management of these funds is through a Green Climate Fund (GCF) established by the Cancún Agreements, which operates under the guidance of, and is accountable to, the COP. The fund is to be governed by a 24-member board with equal representation from developed and developing countries and supported by an independent secretariat. The World Bank was designated as its interim trustee, subject to a review 3 years after the fund begins operations. The modalities of the fund were delegated to a 40-member Transitional Committee (15 members from developed countries and 25 from developing). The initial meeting was scheduled for March 2011 but this was postponed until April 2011. This delay was a concern as recommendations for the fund were to be submitted to the Durban meeting (COP 17) at the end of 2011.

Adaptation

The Cancún Adaptation Framework was established to enhance adaptation efforts by all countries. The Agreements also established a process to help LDCs develop and implement NAPAs. An Adaptation Committee was established to provide technical support to countries, facilitate sharing of information and good practice, as well as providing advice to the COP on adaptation-related matters. The composition and functions of the Adaptation Committee were to be finalised at the Durban meeting.

The Cancún Agreements established a work programme to consider approaches to address loss and damage associated with climate change, particularly for vulnerable developing countries. This will include a climate insurance facility and other options for risk sharing. The recommendations are due to be submitted to COP 18, which will be held at the end of 2012.

Reducing emissions from deforestation and forest degradation

On REDD+[2], the Cancún Agreements affirm that, provided adequate and predictable support is forthcoming, developing countries should aim to slow, halt and reverse forest cover and carbon loss. They encourage developing countries to contribute to mitigation actions in the forest sector by reducing emissions from deforestation and degradation, conserving forest carbon stocks, sustainable forest management and enhancing forest carbon stocks. As part of this objective, developing countries are requested to develop a national strategy or action plan, national forest reference levels or sub-national reference levels as an interim measure, a robust and transparent national forest monitoring system, and a system for providing information on how the safeguards are being addressed throughout implementation. Countries are to follow safeguards ensuring, for instance, the full participation of indigenous peoples, local communities and other stakeholders.

A long-awaited decision on REDD+ was agreed, which gives a signal that the international community is committed to positive incentives, although it postpones clarity on long-term finance for results-based REDD+. The Cancún Agreements called for nations to make recommendations on REDD+ financing at COP 17 in Durban.

Technology development and transfer

The Cancún Agreements established a Technology Mechanism comprised of a Technology Executive Committee and a Climate Technology Centre and Network. The 20-member committee will consist of experts nominated by countries and appointed by the COP. Its roles will include assessing technological needs and issues, recommending actions to promote technology development and transfer, and promoting collaboration among governments, the private sector and others.

The road to Durban

Grubb (2011) highlighted several key areas where progress was urgently needed in the run-up to Durban:

- Financing. Despite broad agreement on financing in Cancún, the global economic recession meant that new sources of finance and governance needed to be realised.

- Despite achievements in negotiating a component of mitigation via REDD, the bigger issue of how to ensure a rapid transition to a global low-carbon economy remained the 'elephant in the room', and without real progress on decarbonisation, what about the issue of 'financing adaptation to the consequences of failure' (Grubb 2011: 1269).
- The role and influence of the EU. Without support from the world's major emitters, the EU is severely constrained due to internal politics and external competitiveness issues.
- The focus on the AWG-LCA as opposed to the AWG-KP when the focus should have remained squarely on 'what next for the Kyoto Protocol'.
- Inclusion of 'facilitative' clauses into the Kyoto Protocol that would allow progressive realisation of emissions reductions, integrating the world's major emitters into a legally binding structure in phases.
- The focus of mitigation on territorialised production of emissions as opposed to emissions associated with (relatively 'fixed') consumption habits.

Within the context of these major issues, even the strong call for action by governments issued by Christiana Figueres, Executive Secretary to UNFCCC, in the run-up to Durban appears necessary but insufficient.

Durban: 'Are we there yet?'

Most parents will remember this question from their children a few minutes after the start of a journey. Although expectations were low, it would be fair to say that some hoped that perhaps the journey to an agreement could be concluded and a picture of what could happen in the future produced. Negotiations were very protracted and 36 hours after the conference was supposed to end, an agreement, of sorts, was reached. Although there was no agreement on a binding agreement for mitigation, the COP did agree the Durban Platform for Enhanced Action (UNFCCC 2011). The Platform establishes the future direction of the climate regime by initiating a new round of negotiations to be concluded by 2015 (an outcome with legal force) that will be in operation by 2020. The text brings all parties, from both the developed and developing world, onto one track, recognises the emissions gap and tries to resolve the difficult conflict between equity and environmental integrity.

In summary, it would be fair to say that the outcomes from Durban were mixed. In answer to the question 'Are we there yet?' we would have to say, 'No, but in about 8 years time we might be able to see where we are going, but don't worry we're heading in the right direction, I think'. In the meantime, it is unlikely that we will see any real efforts to reduce GHG emissions before 2020. Emissions are likely to continue rising, and by 2020 GHG concentrations are likely to exceed 400 ppm and we could well be experiencing more extremes and be close to the point of dangerous climate change. This makes adaptation even more important.

In terms of adaptation, there was some positive news. There was momentum on a number of adaptation-related decisions supporting implementation of the Cancún Adaptation Framework. The Adaptation Committee established in Cancún was operationalised in Durban. Agreement was achieved on membership, authority and modes of work, with developing countries securing a majority of seats on the Committee. The Committee will play a coordinating role, provide advice to various UNFCCC bodies and share information both within and outside of the UNFCCC. The Committee will report to the COP through its subsidiary bodies, making it somewhat less authoritative than if it reported directly to the COP.

A process was agreed to enable LDCs to formulate and implement national adaptation plans (NAPs). Although this is voluntary (as is provision of funding to support NAPs), it seems likely that the UNFCCC Secretariat and the LDC Experts Group will conduct workshops and other activities to provide technical support for the NAPs, and the decision calls for tracking of whether and how developed countries provide financial support. The decision also resolved a contentious matter by creating a process for exploring how non-LDCs can be supported to develop NAPs.

There was agreement to begin a new Work Programme on Loss and Damage and to renew the existing Nairobi Work Programme on various adaptation issues. This means two series of adaptation workshops and technical reports for the coming year. The loss and damage debate has been controversial within the UNFCCC because developed countries prefer to avoid discussions that may link to questions of liability, so a work programme represents a step in a constructive new direction.

Perhaps one of the most important outcomes from Durban was the launch of the GCF. The first meeting of the GCF was held from 23 to 25 August 2012 in Geneva where developed and developing world co-chairs were elected. The fund is meant to be the biggest single funding route for the USD 100 billion that has been pledged to support poor nations each year by 2020 to help them *cut GHG emissions* (author's emphasis) and adapt to the effects of global warming. The GCF, however, will be only one route among many for those funds. Private sector and government cash flowing through other routes will also count towards the USD 100 billion goal. There is a lack of clarity on how long-term finance to support developing countries will be raised and mobilised. There are likely to be very protracted negotiations ahead.

So did Durban live up to expectations? Well that would depend on the level of your expectations. What is interesting is that despite increasing evidence of the impacts of accelerated climate change and increasing variability, there seems to be little political will to address the problem. Climate science is complex and complicated and defies precise forecasts, not least because science itself is based on systematic doubt. The IPCC has laboured to produce evidence for decision-makers to base their decisions upon, but it acknowledges that there are huge uncertainties. Anecdotally, the weird weather being experienced, the increased melt of the Arctic ice sheet, the drought in the United States and its impacts on agriculture, and floods in Asia should be indicators to political leaders that there is a need for urgency. It is clear that there is a conflict between the economic drivers that political leaders have to face, especially since the global credit crunch, and the misplaced belief that it is possible to use economic tools to get the global economy back on track. Climate science, as climate scientists readily admit, is uncertain. Urry (2011) argues that this is part of the problem, the fact that economists present economic theory as 'hard fact', whereas scientists are very quick to precisely communicate areas of uncertainty. But politics often requires the simplification of complex problems so that it can engage society (democracy), sometimes at the expense of shared scientific understanding.

In the next chapter, we discuss the options for decision making in an uncertain world.

3
CLIMATE EXTREMES

Does a post-normal approach make sense?

> We can't solve problems by using the same kind of thinking we used when we created them.
>
> (Albert Einstein (1879–1955))

Introduction

Accelerated climate change and increasing variability is the single largest threat to sustainable development and achieving the MDGs. While the work of the climate community is laudable, the main focus, until recently, has been on mitigation. It is now generally acknowledged that it is unlikely that the international goal of keeping the average global temperature rise below 2 degrees Celsius will be met. Adaptation to both long-term and extreme events is increasingly urgent. Effective risk assessment and management strategies are needed.

Sustainable development is the overall framework for both assessing and managing climate risks. This provides the opportunity to develop an integrated approach to climate adaptation and DRR. Low-regret measures that work within the range of future and projected changes in climate conditions can be used as the starting point to address projected trends in exposure, vulnerability and climate extremes. Such measures have the potential to offer benefits now and lay the foundation for addressing projected changes. Many of these low-regret strategies produce co-benefits, help address other development goals, such as improvements in livelihoods, human well-being and biodiversity conservation, and help minimise the scope for maladaptation.

Potential low-regret measures include early warning systems; risk communication between decision-makers and local citizens; sustainable land management, including land-use planning; and ecosystem management and restoration. Other low-regret measures include improvements to health surveillance, water supply, sanitation, irrigation and drainage systems; climate-proofing of infrastructure; development and enforcement of building codes; and better education and awareness.

Climate risk management

There is considerable uncertainty associated with climate change projections. IPCC projections shown in Figure 3.1 for global temperature rise suggest that for scenarios A2, A1B and B1, temperature will increase by about 1 degree Celsius between 2000 and 2040 (IPCC 2007b).

On the one hand, this can be regarded as comforting because it might appear to some that we have sufficient space in which to make decisions about mitigation; however, research has shown that this can be misleading because there is little 'wiggle room' and decisions about mitigation trajectories and timescales will need to be made soon (Anderson and Bows 2008). The real worry is the political cycle. Politicians, particularly in democratic societies, rarely serve for long periods. In the struggle to either remain in power or be elected to power, it is unlikely that leading politicians will make decisions that will prove difficult for the electorate to accept. They will always promise a better tomorrow: 'not in my term of office'. Governments are fearful of being first. They are caught in what is known as the Giddens' paradox, i.e. governments will not act until something actually goes wrong (Giddens 2009). This may partially explain why there has been little progress in international climate negotiations. No politician will want to make an agreement that undermines the livelihoods of their electorates. President Bush made that absolutely clear when he refused to sign the Kyoto Protocol.

In 2006, Lord Stern and others argued that it is more cost effective to deal with the problem now as opposed to dealing with the consequences later. This seems to have made little difference. The 2008 Credit Crunch has transfixed politicians and, although there was evidence of some slowdown in GHG emissions due to slowing economic growth, the emission trajectory has quickly gone back on track.

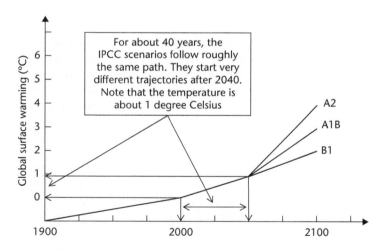

FIGURE 3.1 Temperature increase projection

Source: Adapted from IPCC (2007b).

Recent data show emissions from fossil fuel combustion and other sources, such as gas flaring and cement production, increased by 3 per cent in 2011. Emissions fell in 2008 due to the economic crisis but surged by 5 per cent in 2010, making an average annual increase in emissions of 2.7 per cent over the last decade. The top five emitters are China (29 per cent), United States (16 per cent), EU (11 per cent), India (6 per cent), Russian Federation (5 per cent) and Japan (4 per cent). Global emissions of carbon dioxide (CO_2) increased by 3 per cent in 2011 to 34 billion tonnes. This is the highest level recorded to date.

Since 2000, an estimated total of 420 billion tonnes CO_2 was cumulatively emitted due to human activities (including deforestation). Scientific literature suggests that limiting average global temperature rise to 2 degrees Celsius above pre-industrial levels (the target internationally adopted in UN climate negotiations) is possible if cumulative emissions in the period 2000–2050 do not exceed 1,000 to 1,500 billion tonnes CO_2. If the current global increase in CO_2 emissions continues, cumulative emissions will surpass this total within the next two decades (Oliver *et al.* 2012). It seems unlikely that the Kyoto Protocol targets will be met and any future agreement to reduce GHG emissions will require drastic cuts if the global average temperature rise is to be kept below 2 degrees Celsius. The longer a decision is delayed the more severe the cuts will have to be.

Climate risks

With the huge uncertainties surrounding the mitigation agenda and what might happen if the 2-degrees Celsius limit is crossed, climate adaptation has become an urgent need. It is worth noting that the 2-degrees Celsius limit was first set by the EU in 1996 as a measure that should guide discussions in reducing emission levels. Tol (2007) argues that the target was based on a narrow and uncertain set of climate and economics studies and was somewhat arbitrary. This may be the case, but the 2-degrees Celsius figure does seem to have some sort of resonance and became the benchmark for establishing international targets for emission rates and GHG concentration levels. The general fear is that a rise above the 2-degrees Celsius figure could trigger tipping points and irreversible changes. We do not know.

As the climate warms the weather response will be variable as the system adjusts to a new global average temperature. Recently, we have seen examples of what can be described as odd weather events, for example, in 2012 an unusually dry winter in the United Kingdom was followed by high levels of precipitation during the late spring and early summer months. This was attributed to the jet stream moving further south than usual for that time of year, which brought low pressure systems to the United Kingdom. This southerly movement of the jet stream in Europe was matched by a northerly movement that led to drought conditions in the mid-west and southern states of the United States. There are strong suspicions that this event is linked to climate change, but the mechanisms are not fully understood.

Francis and Vavrus (2012) have demonstrated a link between Arctic Amplification (AA) and extreme weather in mid-latitudes. During the last few decades, the Arctic has warmed at twice the rate compared to the rest of the entire northern hemisphere. The loss of Arctic sea ice allows the exposed seawater to absorb more of the sun's energy during the summer and this is released as the sea begins to freeze during the autumn. In addition, research suggests that warmer air over high latitude land results in earlier snowmelt and drying of land. Both of these effects contribute to conditions that favour the types of extreme weather caused by persistent weather conditions, such as drought, flooding, heat waves and cold spells in the northern hemisphere mid-latitudes.

Can the persistent weather conditions associated with recent severe events, such as the snowy winters of 2009/2010 and 2010/2011 in the eastern United States and Europe, the 2010 European and Russian heat waves, the historic drought and heat-wave in Texas during summer 2011, record-breaking rains in the north-east United States during the summer of 2011, the prolonged winter drought in the United Kingdom followed by storms and flooding, floods and landslides in Seoul in 2011, floods in Beijing and Manila, and the continuing drought in the south-eastern United States in 2012 be attributed to enhanced high-latitude warming? The authors do not claim that this is a nailed-on fact, but research does suggest that human actions are having an impact.

Preliminary results from Cryosat 2, the world's first satellite to be built specifically to study sea-ice thickness, indicate that 900 cubic kilometres of summer sea ice has disappeared from the Arctic Ocean over the past year. This rate of loss is 50 per cent higher than most scenarios outlined by polar scientists and suggests that global warming, triggered by rising GHG emissions, is beginning to have a major impact on the region. In addition, this warming is also releasing methane, a powerful GHG, into the atmosphere. There is a danger that the jet stream will become even more unstable, driving more extreme weather events (McKie 2012). It is becoming increasingly clear that the disturbed weather patterns seen across the globe will become the new normal.

Taking a post-normal approach to climate risk asks questions about the type of risks we will face. We know that making any kind of accurate forecast is impossible, but it is possible to learn from trends – has the weather been cooler or warmer, wetter or drier, do weather events seem more extreme? We can also learn from the experience of those affected by disruptive weather events. There is certainly a wealth of experience from events lately.

A recent study by James Hansen, head of Nasa's Goddard Institute for Space Studies, shows that there has been a sharp increase in the frequency of extremely hot summers. Between 1951 and 1980 these events affected between 0.1 and 0.2 per cent of the world's land surface each year. Now, on average, they affect 10 per cent and the research finds it is highly unlikely that natural variability in the climate system created these extremes. Both the droughts in the Sahel and the United States crop failures are likely to be the result of climate change (Hansen et al. 2012). Given the growing evidence of anthropogenic interference with the climate system, we appear to be on the brink of an era where this global experiment is taking us to conditions we have never experienced. The problem is that if we do not like what we find, there is no going back.

Climate change is a wicked problem because there is little opportunity to learn by trial and error or any exit point from the problem (Ritchey 2007). There are huge uncertainties in climate change, for example climate sensitivity (i.e. the responsiveness of the temperature of the climate system to a change in radiative forcing) and the likelihood of abrupt non-linear events. It is likely that the uncertainties in probabilities and consequences may not be resolved with a high degree of certainty before we have to think about dealing with the implications. The anthropogenic production of GHGs by, primarily, the industrialised countries, is generating a new family of produced unknowns. Although known in general terms, the 'what' and the 'when' cannot be predicted with any accuracy (O'Brien *et al.* 2010), but, as Hansen points out, they are getting more frequent (Hansen *et al.* 2012). There are two adaptation problems: adjustment to long-term change and adjustment to increasing variability and extremes. There will be multiple hazards, for example high rates of precipitation can cause both flooding and landslides, and increased temperatures can lead to droughts, which will impact food security and human and livestock health.

Perceptions of climate risks

As well as the huge uncertainties surrounding climate change, there are many contradictory views. There are climate deniers who do not accept that human activities are changing the climate (see Box 3.1), there are those who accept climate change but do not believe that it is a serious problem, and others who believe that it may be too late to deal with the problem. Climate science has been unable to provide any real certainty and the further we have tried to look into the future the more difficult it has been to provide any reliable predictions about what might happen. Conventional scientific methods cannot provide sufficient data for robust policy making – climate change uncertainties and the interactivity between systems at micro, meso and macro scales militate against gathering enough empirical data for robust decision-making. A different approach to assessing outcomes is needed. Funtowicz and Ravetz (1991) devised a methodology of inquiry, termed postnormal science (PNS), that deals with issues such as global environmental change where facts are uncertain, values in dispute, stakes high and decisions urgent. PNS provides a lens that can be used to aid decision making.

BOX 3.1

CLIMATE DENIERS

Deniers actively claim that climate change is a myth or hoax, something promulgated by climate scientists and those who want to rule the world. This is not a constructive or useful way of challenging climate science. Unlike the scientific community, which puts its findings into the public domain for scrutiny, the deniers are a secretive group, particularly about who funds them. Unsurprisingly, they demand more and more data from the climate community. In some cases, they actively harass climate scientists. Take the case of Michael Mann, one of the scientists responsible for the so-called 'hockey stick' graph. In a recent interview with *The Guardian* on his forthcoming book, *The hockey stick and the climate wars: Dispatches from the front lines*, he states that he has been regularly vilified on Fox news, contrarian blogs and by Republican members of the US Congress and now accepts that refuting deniers is part and parcel of the job he does (Goldenberg 2012).

The Center for Responsive Politics' nonpartisan, independent and non-profit research group that tracks money in US politics, highlights on their website[1] the contributions made to both Democratic and Republican politicians and parties received from the major energy companies, some USD 39.5 million in 2012. The Republican Party has received over USD 27 million (Opensecrets.org 2012).

A recent study by the Berkeley Earth Group, partly funded by climate sceptics (USD 150,000 from the Charles G. Koch Charitable Foundation, set up by the US billionaire coal magnate and key backer of the climate-sceptic Heartland Institute think-tank), agrees with climate change modellers that there is a discernable increase in temperature caused by human actions (Berkley Earth Temperature Group 2011). Professor Muller, a physicist and climate change sceptic who founded the Berkley Earth Temperature Group (Best) project, said he was surprised by the findings and now has changed his mind about climate change (Hickman 2012). It would be interesting to ask Charles G. Koch if he believes that his money was well spent and if his view has also changed.

Has this affected public opinion? In the EU, a survey by Eurobarometer found that just over half (51 per cent) of respondents consider climate change to be one of the world's most serious problems (and 20 per cent feel it is the single most serious problem). Overall, it was seen as the second most serious issue facing the world, after poverty, hunger and lack of drinking water, and a more serious problem than the economic situation (Eurobarometer 2011). In the United States, an analysis of Gallup surveys found that in 2004 26 per cent of respondents said they worried 'a great deal' about climate change; in 2007 that number rose to 41 per cent; by 2010, it had fallen to 28 per cent. This variation exists despite consensus among scientists about the underlying data patterns and virtual unanimity of scientific opinion (Brulle *et al.* 2012).

There is clearly a difference in opinion between Europe and the United States, and this could be attributed to the better funded and organised denial campaign. It is noteworthy that none of the candidates in the 2012 US presidential campaign made climate change an election issue, despite the severe drought in the US farm-belt.

> The climate denial campaign is not confined to the United States. In the United Kingdom, Viscount Monckton of Brenchley, who wrongly claims to be a member of the House of Lords (the upper chamber of the UK government), has peddled a long line of nonsense on climate change science despite his claims being refuted on numerous occasions. For example, John Abrahams, who attended a speech given by the Viscount, was so incensed that he wrote to all of the scientists and bodies that Monckton cited and found that their work had been wholly misrepresented (Abrahams 2010).
>
> Australia is a coal superpower and the world's largest exporter of this dirtiest of all fossil fuels. Australia is also very vulnerable to climate change, with projections of longer droughts and bigger floods. In Western Australia, rainfall has already declined by 20 per cent. The 2010–2011 Queensland floods claimed 35 lives, affected 70 towns and 200,000 people, and was estimated to have cost the economy some AUS$30 billion. In 2009, the tragic 'Black Saturday' bush fires, exacerbated by an unprecedented heatwave, claimed 173 lives. Viscount Monckton of Brenchley conducted a 3-week lecture tour in Australia in the summer of 2011, sponsored by Hancock Prospecting, a mining company with interests in coal, and organised by the Climate Sceptics party. This party, established in 2009, claims to be the first political party in the world committed to exposing the fallacy of human-driven climate change (Lewandowsky 2011).
>
> Perhaps the above might help to explain why it has been so difficult to make progress in reducing GHG emission.

So why are views on climate change so polarised? As we have seen in Box 3.1, views on climate change in the United States have shifted between 2002 and 2010 (Brulle *et al.* 2012). Brulle *et al.*'s analysis suggests that elite cues and structural economic factors have the largest effect on the level of public concern about climate change. Elite cues are generated by elite conflicts over climate change, for example differences between political parties, and individuals tend to support the views expressed by an elite group they subscribe to. Structural economic factors refer to issues such as the state of the economy, energy prices and unemployment – in difficult times factors such as these are more likely to have greater significance. Furthermore, their analysis found that while media coverage exerts an important influence, this coverage is itself largely a function of elite cues and economic factors. Weather extremes have no effect on aggregate public opinion. Promulgation of scientific information on climate change to the public has a minimal effect. This implies that information-based science advocacy has only had a minor effect on public concern, while political mobilisation by elites and advocacy groups is critical in influencing climate change concern.

It would seem reasonable to expect that extreme weather events would have a significant influence on public perceptions. Research by Goebbert *et al.* (2012) into the American perceptions of weather change finds that this is not necessarily the case. Actual weather changes are less predictive of perceived changes in local temperatures, but better predictors of perceived flooding and droughts. The research finds that cultural biases and political ideology also shape perceptions of changes in local weather. In short, beliefs about changes in local temperatures have been more heavily politicised than is true for beliefs about local precipitation patterns.

We can conclude from this that despite the wealth of scientific evidence showing that we are changing the climate and evidence of 'weird' weather patterns, the way this is addressed by political and community leaders and the way in which the media feeds those views to the wider public is more important than the evidence. It might be easy to conclude that the climate denial campaign has been effective. Clearly the energy giants have funded the climate denial campaign not because they have evidence that definitely shows that climate change is driven by factors other than anthropogenic activities, but because they have a vested interest. In an interview with Al Jazeera, Bill McKibben (Schumann Distinguished Scholar at Middlebury College, founder of the global climate campaign 350.org and the author of *Earth: making a life on a tough new planet*) estimates the value of assets held by energy giants and energy producing states at some USD 20 trillion. If they subscribed to the anthropogenic view of climate change, then these assets would effectively be worthless. For large global energy conglomerates and energy-producing nations, this would spell financial disaster (McKibben 2012). Though some might think that this would be desirable from an ecological perspective, there are many others who would think it would be lunacy. The status quo interpretation of sustainable development is the dominant view.

Why is this the case? Governments and large conglomerates act from perceived self-interest. Powerful messages from the media supporting the status quo clearly have an influence, as well as the recognition that to shift to a low carbon trajectory would be challenging. Neo-liberalism, with its misplaced belief in the rationality of market choices, dominates our consumer-based lifestyles. Our reflexive world would suggest that we would be very sceptical of political views peddled through the press, either in support of action to tackle climate or denying it is an issue, but as Kahan points out:

> People with different values ... individualists compared to egalitarians, for example ... draw different inferences from the same evidence. Present them with a PhD scientist who is a member of the US National Academy of Sciences, for example, and they will disagree on whether he really is an 'expert', depending on whether his view matches the dominant view of their cultural group.
>
> (Kahan 2012: 255)

Kahan argues that individual positions on climate change have come to signify the kind of person one is. An individual whose beliefs differ from those with whom they share their basic cultural commitments risks being labelled as weird and obnoxious in the eyes of those on whom they depend for social and financial support. In short, do not stand out from the crowd!

Many have been bludgeoned into accepting the status quo because it is easier to believe that incremental changes rather than wholesale shifts will get us out of the problems we face, even if they doubt the validity of the science. This is very much status quo thinking and despite scepticism, it is easier to deal with 'the devil you know' as opposed to the one you do not. It is a sort of insurance approach. But as Kahan suggests we need a shift in cultural values, a new belief system, one that values human well-being as opposed to consumer choice.

But there are longer-term consequences. As we all know, it is impossible to predict the future and we use scenarios for almost all major planning. Scenarios are a powerful strategic tool used by a diverse multitude of organisations to systematically ascertain the most likely future outcome for any given situation in order to assist decision making and planning by identifying problems, threats and/or opportunities (Bradfield et al. 2005). Essentially, scenario planning is the amalgamation of several rational futures selected from a number of probabilistic outcomes that are developed in enough detail to be highly plausible, using extrapolative, prospective and/or reductionist approaches (Kelly et al. 2004).

Generally, scenarios fall into two broad categories: exploratory or normative. Explorative or descriptive scenarios typically centre on the outcome of decision making with conceivable endpoints. These scenarios can be defined as a narrative examination of a consistent chain of factors including trends, events or probable outcomes most likely to follow from existing actions (forecasting) including policies, attitudes and behaviour. Normative or strategic scenarios encompass the rationale of commencing with a desired outcome and working backwards, termed 'backcasting' or, as a reference case, to arrive at the most sustainable scenario or desirable future that also accounts for a consistent suite of trends and/or patterns (O'Brien et al. 2007; Kelly et al. 2004; IEA 2003; Schwartz 1991; Huss 1988).

Successful fruition of scenarios, however, does not necessarily mean a wholly accurate depiction of future events is ultimately achieved; rather, establishment of scenarios provides a vehicle to enhance decision making by extrapolating probabilistic factors to shape a range of alternative future conditions, therein equipping planners and policymakers with the knowledge to mitigate or adapt to future conditions. Schwartz (1991) reminds us that scenarios are not predictions, but simply vehicles for helping people learn.

As stated, scenarios are used by a range of organisations for a variety of purposes. The Tellus Institute, based in Boston, United States, has developed a suite of scenarios entitled 'The Great Transition Initiative' (GTI), which elaborates scenarios for global futures (Raskin et al. 2002). They point out that there are a number of possible futures, which depend largely on the interpretation of risks faced and the responses fashioned. Their analysis starts by stating that human society is at a watershed in development (a post-normal era), many futures are possible and it is the decisions taken now that will influence the shape of planetary society. The analysis offers three classes of scenarios:

1. The *Conventional World* scenario where it is assumed that incremental changes to markets and policy adjustments can cope with social, economic and environmental change.
2. The *Barbarisation* scenario assumes that change cannot be managed incrementally, resulting in either *Breakdown* or *Fortress World*:
    a. *Breakdown* sees crises spiral out of control, institutional collapse and the world descending into anarchy and tyranny.
    b. The *Fortress World* sees the emergence of authoritarian regimes and the development of protected enclaves.
3. The final scenario, *Great Transitions*, envisages the emergence of new values and development paradigms that emphasise quality of life and material sufficiency, human solidarity and global equity, and affinity with nature and environmental sustainability.

Following on from this, the group has established a Global Citizens Movement (GCM) and a group known as The Widening Circle (TWC). GCM refers to a profound shift in values among an aware and engaged citizenry. Although governments, multinationals and NGOs are powerful actors, they are nonetheless influenced by millions of citizens throughout the world. There is a widespread latent desire among concerned citizens who recognise that the world must address a suite of deepening social, economic and environmental problems, but do not yet know how to take action themselves (Kriegman 2007). TWC aims to strengthen the GCM – it is the action arm of the GCM. The purpose of these movements is to explore new forms of governance.

Though nation states are powerful, problems, such as climate change, can only be addressed through transnational action, hence the development of the climate convention. There are, however, issues concerning the distribution of power that need to be addressed. As debated in this book, the developed world has effectively kept the debate at status quo. Arguably, this group has been strongly influenced by what can be termed 'shadow states' – a group of international actors that move resources and finance around the world in order to maximise profits. This sort of activity is largely unregulated and it is able to use its financial muscle to influence decision makers. Global environmental governance is arguably well established, but not necessarily effective. This makes it problematic to consider what kind of governance is needed for national and sub-national adaptation strategies.

Climate science

In what is termed 'normal science', method proceeds through experiment, data collection, data analysis and evaluation within an established paradigm (Kuhn 1962). This is not to suggest that science does not progress. Kuhn argues that scientific knowledge progresses through socially constructed paradigm shifts, where normal science is what most scientists do all the time and what all scientists do most of the time. The process of paradigm shifts is as follows:

1. from normal science (the rules are agreed or disagreed in debates that cannot be concluded; science is puzzle solving, but some contradictions in theory cannot be resolved);
2. to revolutionary science (important rules are called into question; contradictions may be resolved; paradigms shift);
3. to new normal science (new rules are accepted; science returns to puzzle solving under new rules).

Science is an unfolding process of discovery, but as we move outwards from the applied science field, both uncertainties and stakes increase (Figure 3.2). The professional consultancy field is an area of expert advice either from specialists in the field or those with considerable experience. In certain situations even this may not be sufficient to eliminate uncertainty, for example the surgeon who has investigated the condition of the patient as thoroughly as possible may only be able to resolve uncertainties after the first incision. Of course, time is of real importance to the patient because he/she cannot afford to wait until the condition has been categorically proved, particularly if it is life threatening. As we move to the post-normal science field, uncertainties increase but the pressure from stakeholders for answers also increases.

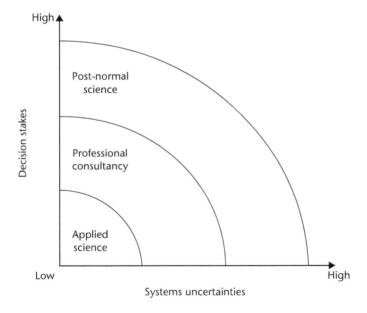

FIGURE 3.2 Post-normal science
Source: Funtowicz and Ravetz (1991).

There are, however, occasions when some stakeholders, such as climate policymakers, want or need to know the answers well before normal science has resolved the deep inherent uncertainties surrounding the problem at hand. Typically, there are two options for dealing with uncertainty:

1. placing boundaries on the uncertainty; and
2. reducing the effects of uncertainty.

The first option involves normal scientific study, i.e. reducing the uncertainty through data collection, research, modelling, simulation techniques and so on, but the huge uncertainties of climate change mean it is unlikely that we can resolve these issues in a suitable timescale for policymakers. Policymakers will therefore have to revert to the second option – to manage uncertainty by integrating uncertainty directly into policymaking.

Is a post-normal approach to risk management and climate adaptation justified?

If, as Funtowicz and Ravetz (1991) claim, we have entered an era where post-normal science provides an appropriate and effective vehicle for addressing the many complex problems we face, then this suggests strongly that we live in a post-normal world. Is this so? It is certainly true that accelerated climate change is driving the world to a set of conditions that humans have never experienced. The world today appears to be in a transition period when the old orthodoxies are dying but the new ones have not yet emerged (Sardar 2009). Sardar identifies three characteristics of this transition period: complexity, chaos and contradictions. There is little doubt that global change driven by climate change, a growing and urbanising global population, increasing resource scarcity, chaos in the global financial system, the shift in economic power to the east and political movements such as the Arab Spring, signal that a new era is beckoning. But what that era will look like is very uncertain.

The complexity of today's world is clearly illustrated by the global energy problem. There is a clear need to shift to a low carbon pathway; however, as O'Keefe et al. (2010) argue, there are contradictory signals from the climate community, governments and business as well as large uncertainties about supplies and geopolitical uncertainty about access. This is especially true for areas with limited indigenous energy resources, such as Europe, which is increasingly vulnerable to global uncertainties. O'Brien (2009) points out that the steps taken within the EU are likely to increase vulnerability. The EU's plans to marketise the energy system will benefit shareholders, not customers, and slow investment in new low carbon capacity. Plans to forge agreements with energy producers will continue the reliance on imports and while much has been said about the benefits of efficiency there is little evidence of concrete actions. There is a new urgency in developing unconventional hydrocarbons (tar sands and shale) and renewed interest in nuclear technology. Though progress is being made in renewable technologies, the sheer-scale of the hydrocarbon business sector means that it can effectively dictate energy policy. Hydrocarbons look set to dominate the world energy scene for some time to come, despite clear evidence of their adverse effect on the global climate. In short, the global energy picture is highly complex and interconnected. Failure is a characteristic of highly interconnected complex systems (Perrow 1999).

Our networked global society is highly vulnerable to technical failures, viruses and hackers that cause chaos. Stock markets can be chaotic, driven by computer trading where their programmes respond to small changes regardless of the circumstances. But chaos can also be driven by social networks, such as Facebook and Twitter, and mobile phones. The UK 2000 fuel blockade started as a simple demonstration but the use of radios and mobile phones quickly spread the protest throughout the United Kingdom and led to the government capitulating to the protestors' demands (O'Brien 2006). The 24-hour media world we live in and the ability to disseminate images and messages spreads information around the globe. The Arab Spring has been driven by a burning desire for freedom but those fires have been fanned by social networks. The posting of phone camera videos on YouTube or blogs have been quickly picked up by the media and, despite the ever-present claim that the images and sounds could not be independently verified, they have nonetheless influenced public opinion and consequently political leaders.

Both complexity and chaos lead to contradictions. Our world appears to be changing rapidly with ever-faster communication and faster technological innovation, such as super broadband and smaller and more powerful processors. But the more things change, the more some things stay the same. War and conflict still continue and the vast majority of the people in the world are still hungry. Although there has been an exponential increase in knowledge, there are areas where we are still ignorant, for example many non-Muslims understand little about Islam. This form of ignorance can lead to fear and prejudice. There are other contradictions where we introduce new uncertainties, for example we do not know what the effects of nanotechnology or genetic modification of crops will be; it may be a number of years before we have sufficient knowledge. But there is a positive side to contradictions: they provide us with a perspective that prevents oversimplified analysis of problems or situations. We are forced to consider clashing trends, viewpoints, facts, hypotheses and theories, and realise that the world is not amenable to naïve one-dimensional solutions.

Complexity, chaos and contradictions feed uncertainty. Climate change is complex and the range of voices in the debate and the often contradictory claims by climate change believers, sceptics and deniers produces a very noisy environment in which policymakers have to try and balance conflicting interests in their deliberations. It is no surprise that little progress has been made. In the meantime, those least responsible for the problem are suffering the most. Accelerated climate change and increasing variability are causing some 300,000 deaths a year and are directly affecting 325 million people in the least developed countries. Economic losses are projected at USD 125 billion a year. Over half of the world's poor are vulnerable and some 500 million are at extreme risk from weather-related disasters that bring hunger, disease, poverty and lost livelihoods (Global Humanitarian Forum 2009).

Sardar (2009) urges a rethink of our cherished notions of progress, modernisation and efficiency. Our model of progress is based on continuous economic expansion. This fails to take into account the ecological limits of the planet. We really do not yet understand what those limits are because, historically, technological innovations, such as improved agriculture, have allowed continued expansion. This belief in our capacity for innovation is very much part of a weak interpretation of sustainability, but if this direction of travel continues, there is a danger that we may reach those ecological limits, perhaps unknowingly, driven by a neo-liberal capitalist system that in the end will devour itself. As Sardar points out, there is need for change based on the virtues of humility, modesty and accountability, and the indispensable requirements of living with uncertainty, complexity and ignorance. Climate change may well be the vector that shifts the current trajectory.

From this perspective, it is clear that a post-normal perspective is a very different lens through which to view the world. In general, in normal science the role of the scientist is to describe the world, not as it is experienced, but as it is 'in reality', independent of experience. This is termed 'scientific realism', which posits scientific theories as 'true' or 'approximately true' descriptions of the world in the sense that the entities and processes set out in theory have some relationship to entities and processes in the external world. Scientific realism claims that as science makes progress, i.e. as scientific theories successively improve, they are able to answer more and more questions. In summary, three features of scientific realism can be discerned:

1. it describes the world as it is;
2. it describes scientific theories as 'true' or 'approximately true' descriptions of the world; and
3. at times is warranted in judging scientific theories to be 'true' or 'approximately true'.

Many would be comfortable with the above for disciplines such as physics or chemistry, but for social sciences, such as psychology and sociology, some, or all, of the above may not be true. Physical sciences, or in our case climate science, deal in facts or truth to provide an evidence base for policy making, whereas social sciences are value laden. Post-normal science takes this further by arguing that a range of stakeholders, including scientists, politicians, businesses, the public sector and communities, should be part of the decision-making process when there are uncertainties. This raises issues of governance in adaptation.

What Funtowicz and Ravetz (1991) have captured is that evolution of any kind is not a linear or gradualist process. Work by Eldredge and Gould (1972) on the evolution of species disputed the idea of gradualist evolution. In short, they showed that evolutionary change is punctuated by events that can drive evolution in new ways. The punctuations that Funtowicz and Ravetz refer to are phenomena that cannot be addressed by conventional methods. In our case, accelerated climate change and increasing variability have driven a need for a new way to think about climate problems so that we can devise solutions. In short, change, whether this is in thinking or doing, is shaped by events or punctuations that require us to do things in different ways. Later in this book, we discuss the importance of learning processes in helping us respond to challenges. In many respects, this is the normal way in which we have evolved, and today we find ourselves in a situation where we are confronted by change and a need to elaborate new ways of dealing with a novel situation.

Given the uncertain nature of future climate and society, the range of perspectives on uncertainty, and the fact that adaptation as rational decision making is not working (Elahi 2011; Inderberg and Eikeland 2009), PNS can be employed as a heuristic lens to link epistemology and governance with a focus on 'wicked problems', such as climate change (Farrell 2011; Kastenhofer 2011):

> ...wicked problems not only ask for a post-normal science or socially robust knowledge ... but even more so for a governance regime that is robust for and can make sense of post-normal approaches to knowledge production, validation, and application.
>
> (Kastenhofer 2011: 309)

Wicked problems are not simple to define because they are complex issues with non-linear causality with solutions that are 'better or worse', not 'right or wrong', because they involve value judgements (Brown *et al.* 2010; O'Brien *et al.* 2008). Petersen *et al.* (2011: 365) identify three key elements of PNS:

1. The management of uncertainty: PNS acknowledges that uncertainty is more than a technical number or methodological issue. Ambiguous knowledge assumptions and ignorance give rise to epistemological uncertainty.
2. The management of a plurality of perspectives within and without science: complex problem solving requires scientific teamwork within an interdisciplinary group and joint efforts by specialists from the scientific community and from business, politics and society.
3. The internal and external extension of a peer community: an extended peer community includes representatives from social, political, and economic domains that openly discuss various dimensions of risks and their implications for all stakeholders.

This tells us that there needs to be an open and participatory approach to thinking about climate risks; no longer is a purely expert or professional approach plausible. Placing climate change risks into a wider domain of stakeholders means a more democratic decision-taking process. There is no guarantee that we will get it right every time, especially given the uncertainties of climate change and the fact that we have little idea of what the global average temperature will rise do, but it does seem that the 2 degrees Celsius target is unlikely. As Urry points out:

> Human and physical systems exist in states of dynamic tension and are especially vulnerable to dynamic instabilities. Various systems may come to reverberate against each other and generate larger systemic impacts. It is the simultaneity of converging shifts that creates significant change. Thus resource depletion, climate change and other processes may overload a fragile global ordering, creating the possibility of catastrophic failure.
>
> (Urry 2011: 44–45)

The main points of contention with a PNS approach to climate change are:

- the ever present need to avoid privileging some 'truths' over 'others'; a democratic approach should help to place knowledge in an appropriate context; and
- that decision makers rarely make rationally optimal decisions based on science, whether that science be traditional or post-normal (Farrell 2011; Petersen *et al.* 2011; Inderberg and Eikeland 2009).

This has strong implications for governance. Evaluating evidence and new knowledge in a democratic forum implies some sort of risk sharing in the decision-making process itself. Provided that decisions are made within a 'low-regrets' context and deal with real problems, as opposed to thinking about large solutions that are only likely to deal with longer term problems, then neither of these concerns mean we should avoid a fuller, more inclusive and transparent dialogue on the actual uncertainties of climate risk. Epistemological limits to climate prediction should not be interpreted as a limit to adaptation, despite the widespread belief that it is. Without the science we cannot make rational decisions. By avoiding placing climate prediction and consequent risk assessment at the fore, it should be possible to develop successful adaptation strategies despite the deep uncertainties of climate change (Dessai *et al.* 2009; Sheppard *et al.* 2011; Tebes 2005). Normal climate prediction and risk assessment are not suitable vectors for adaptation, but this is no reason for slowing down or stopping research effort in these fields.

Post-normal risk management

As the climate problem is a post-normal science issue, it follows that related aspects, such as managing climate risks and adaptation, are also post-normal issues. Risk is a function of probability (how likely is an event) and consequences (what is the likely result). Conventional risk assessment makes use of empirical data, for example insurance companies can use accident data to assess levels of insurance for motor vehicles – less experienced drivers with powerful vehicles are more likely to have an accident. But for climate change we have limited data on which to draw. Essentially, we will be making subjective decisions about the nature of the risks we will have to adapt to. This does not preclude expert driven assessment, but it does mean that this

Climate extremes 67

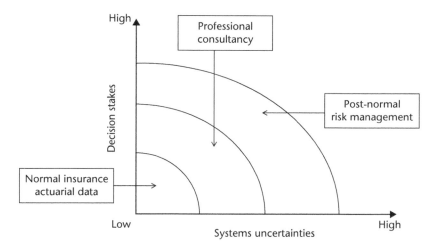

FIGURE 3.3 Post-normal risk management
Source: Funtowicz and Ravetz (1991).

will also be subjective. Drawing on Funtowicz and Ravetz (1991), we can postulate a post-normal risk management view (Figure 3.3).

Figure 3.3 shows that with increasing uncertainties and higher stakes, a new approach to risk is needed. For effective climate risk management, this means a range of actions to both reduce and transfer risk. To be successful, integrated approaches will require a combination of both hard and soft measures (including individual and institutional capacity-building and ecosystem-based responses), but they must be informed by indigenous knowledge and tailored to specific local circumstances. A good example is the work being done following landslides in Seoul in 2011 (Box 3.2).

BOX 3.2

FLOODS AND LANDSLIDES IN THE REPUBLIC OF KOREA

In July 2011, rains and thunderstorms delivered over 495 mm (19.5 in) of rain in the Seoul region during a 2-day span, the heaviest such event in July since 1907. After 3 days, 587 mm (23.1 in) of rain was recorded in the area.

On 26 July, a landslide buried three hotels in Chuncheon, some 50 km east of Seoul, killing 13 people from Inha University. A landslide in Umyeon-dong killed 18 residents in an apartment block. Floodwaters inundated highways and tracks of the Seoul Metropolitan Subway, while bridges over the Han River were closed. Damages have run to hundreds of millions of US dollars. Motor vehicle damages alone reached USD 38 million on 28 July. Nearly 978 ha (2,420 acres) of agricultural land and more than 10,000 homes were flooded. In total, 57 people died in the floods and landslides.

> Following these devastating events, the President of Korea announced that there would be a full-scale review of disaster management and prevention with a focus on climate change threats and improvements to urban-disaster prevention. To undertake this review, a task force was established under the leadership of the Prime Minister's office. The task force was made up of representatives from 13 government departments, representatives from the local governments of Seoul and Gyeonggi-do and five experts. The first meeting was held on 10 August 2011.
>
> The task force met at least 20 times, gathered information from experts and conducted seminars and public hearings in order to collect a broad range of opinions and information from a range of stakeholders.
>
> The task force developed a number of recommendations including:
>
> 1. Landslide protection dams should be built in appropriate locations, which would mitigate the impacts of landslides. To site these dams, a combination of historical data, local knowledge, GIS surveys and climate projections were recommended. The task force estimated that some 10,000 dams would have to be built over the next 10 years.
> 2. An early warning system using sensor technology should be deployed and linked to disaster-response bodies, as well as local government and local people.
> 3. Training exercises should be conducted in order to improve preparedness.
> 4. Participatory approaches to work with local communities should be employed to both identify and manage potentially dangerous landslide areas. This fine-grained approach would strengthen the landslide database and improve community cohesion.
> 5. Different tree and bush planting strategies should be developed that would use deeper-rooted species in order to stabilise slopes.
>
> The recommendations use a combination of hard measures (rock dams, tree planting) and soft measures (early warning system, training exercises) along with local knowledge by users of the area to identify and manage landslide risks.

The multiple facets of climate change hazards, impacts and adaptation strategies, as well as the uncertainty associated with each, preclude a single approach that will be effective in all instances (Yohe and Leichenko 2010). This implies that as we learn more about more about different aspects of climate change, we will need to re-evaluate approaches. An effective climate change policy is an iterative one that considers and incorporates new learning at regular intervals (Figure 3.4).

In Figure 3.4, which has been adapted from the Thames Estuary 2100 Plan, we can see that this iterative approach aims to keep the flexible adaptation pathway with mitigation [5] below the level of acceptable risk [1]. Acceptable risk is usually

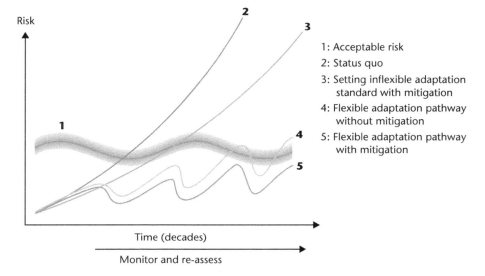

FIGURE 3.4 Flexible adaptation and mitigation pathways
Source: Adapted from Yohe and Leichenko (2010).

set at the national levels and will vary over time as social, economic and technological changes modify attitudes towards risk. Examples include smoking bans, the introduction of seat belts and, more recently, obesity in some countries where new learning has persuaded governments to introduce new norms. But we have little idea of what constitutes an acceptable level of climate risk. Some efforts have been made, for example the introduction of a Heat Wave Plan for the United Kingdom driven by the 2003 heat wave, but the reality is that status quo thinking dominates the climate debate, and pathway 2 in Figure 3.4 is the likely trajectory for GHGs. There is a danger that adaptation thinking will also be mired in status quo thinking, as shown in pathway 3, leading to an inflexible approach to adaptation options. Ideally, we should follow a flexible adaptation pathway with mitigation [5]. The next best option is a flexible adaptation pathway without mitigation [4]. Again, there is a danger that such a path could exceed acceptable risk levels.

In its Fourth Assessment Report, the IPCC makes clear that both mitigation and adaptation should be subjected to an iterative risk management process that takes into account climate change damages, co-benefits, sustainability, equity and attitudes to risk (IPCC 2007b). This is analogous to taking a post-normal approach to climate risk, as shown in Figure 3.3.

Multi-hazard risk management approaches provide opportunities to reduce complex and compound hazards. Considering multiple types of hazards reduces the likelihood that risk reduction efforts targeting one type of hazard will increase exposure and vulnerability to other hazards in the present and future.

The IPCC (2007b) has set out a number of areas where fragmentation between different stakeholders is preventing progress on DRR and climate change adaptation:

Finance

Financing for DRR is low when compared to spending on international humanitarian responses. This does not mean that it is misguided, but it does mean that opportunities for international finance for disaster risk management and adaptation to climate change to create synergies have not yet been fully realised.

Technology transfer

This is an important area because we need to know what works and why, but there is a lack of coordination on technology transfer and cooperation between the DRR and climate change adaptation communities, which has led to fragmented implementation.

Integration across scales

Efforts at the international level do not necessarily lead to substantive and rapid results at the local level. There is a need to improve integration from the international to the local. Integration of local knowledge with additional scientific and technical knowledge can improve DRR and climate change adaptation. Local memory of extreme events can help to build community capacity as well as identify shortcomings.

Participation

Local participation supports community-based adaptation, which brings benefits to climate risk management. It should be noted, however, that community-based adaptation does benefit from the availability of human and financial capital as well localised climate information.

Communication

Risk communication must be timely and appropriate for effective adaptation and disaster risk management. This can be strengthened by explicitly characterising uncertainty and complexity. To be effective, risk communication should build on exchanging, sharing and integrating knowledge on climate-related risks among all stakeholder groups. Among individual stakeholders and groups, perceptions of risk are driven by psychological and cultural factors, values and beliefs.

An iterative process of monitoring, research, evaluation, learning and innovation can reduce disaster risk and promote adaptive management in the context of climate extremes.

In summary, we argue that we need to take a very different approach to dealing with climate change. Whether or not policymakers accept uncertainty in climate change, it is something they will just have to deal with. They do, of course, have strategies in place to deal with complex problems, but the difference is that, in the case of climate change, the problem is global, the science uncertain and the problem is wicked. If they act, then they may lose competitive economic advantages; something that many politicians will shy away from. If they ignore the problem, the consequences could be catastrophic. Many either dislike or are indifferent to politicians, but at least we can recognise the quandary they are in. It would be naïve to expect political leaders to make decisions that would make them unpopular, even if, deep down, we think that is what is needed. The change has to come from us. It will take something extraordinary before public opinion will shift away from its love affair with consumerism and those members of the public, a global minority, who enjoy its benefits. For the moment, it is that global minority controlling many of the global institutions that are, to some extent, a part of the problem. It seems unlikely that the global minority is ready to let go of the levers of power.

4
DISASTER MANAGEMENT

> There are two big forces at work, external and internal. We have very little control over external forces such as tornadoes, earthquakes, floods, disasters, illness and pain. What really matters is the internal force. How do I respond to those disasters?
>
> (Leo Buscaglia (aka Dr Love) (1924–1998))

Introduction

Throughout history people have developed ways of responding to disruptive events. People have learned from adverse experiences and used that learning to better prepare for the future. Today, many countries have dedicated disaster management platforms, but often they are reactive. There are efforts through UNISDR to move to a more proactive stance. There is also recognition that there is considerable overlap between DRR and CCA. This chapter looks at the evolution of disaster management and areas of overlap with CCA.

The evolution of disaster management

Disasters are no longer thought of as having divine origin or intent ('acts of God'), but are firmly rooted in the relationship between people and their environments. There are two broad types of disaster management. The first is what can be termed conventional disaster or emergency management,[1] i.e. those systems established by governments to deal with disruptive events. The second is humanitarian responses, which can be characterised as interventions largely, but not exclusively, by the international community to disastrous events or conflict situations. Typically, these efforts are directed to poorer countries that lack capacity or resources.

When considering the evolution of disaster management, there are broadly two areas to discuss. The first is the evolution of disaster research. The second is the changes in disaster management practice. Both are important if we are to understand the appropriateness of disaster management strategies and techniques for CCA, particularly for extreme events. It is in the post-Second World War era that there have been major shifts in the ways disasters and risks are viewed and the ways in which disaster response systems have evolved. Simultaneously, events themselves and socio-economic and demographic changes have influenced response systems.

Conceptualising disaster management

In terms of disaster thinking, Alexander (1993) identifies six approaches to disaster research (geographical, anthropological, sociological, developmental, medical and technical) but the dominant disciplines, particularly post-Second World War, are geographical and sociological. The geographical approach focuses on human–environment interactions, whereas the sociological approach has as its premise that disasters are social events that reflect the ways we live and structure our societies and communities.

The sociological research, led by Dynes and Quarantelli, has largely been focused on the developed world and addresses essentially the problem of response through an analysis of collective organisational behaviour. It usefully criticises natural hazards work, addressed later in this chapter, as having too much of an emphasis on the individual cause of disaster, e.g. fire, earthquake, flood, etc. (Quarantelli 1992). Critics of the sociological school note that its focus on organisation is primarily to improve the 'command-and-control' system in response mode. Most acutely, Hewitt (1998) notes that such an authoritarian outcome is really addressing social problems without social content.

There is clear recognition that the source of disasters is rooted in the relationship between people and their environments. Much of this viewpoint is established in the natural hazards tradition of geography, but it requires a wider understanding of environmental geography to locate the critical debates.

Smith (1984) brought together the physical and social worlds through his commentary on uneven nature development that linked modes of production to the production of nature and, thus, space. This construction of space builds on Zimmerman's (1933) view of resources, where resources do not exist but become. That form of becoming varies from mode of production to mode of production, for example steel not flint under capitalism for tool making, buildings not caves for dwellings; however, the construction of space and the consequent becoming of resources produce new risks.

This can be explored further. Communities moving to the new urban areas generated by the industrial revolution, or those in the developing world experiencing similar urbanisation today, moved from an environment where risk was broadly related to the 'vagaries of nature', e.g. crop failure, to an environment where new and additional risks were produced, e.g. chemical explosion. By additional, we mean that there is a different relationship with risk generated by the 'vagaries of nature', not the elimination of risk. In that sense, risk chains are created in a context of increasing material development. This is shown in Figure 4.1.

In geography, the natural hazards paradigm was initially centred around the leadership of Gilbert White at the University of Chicago in the late 1950s and early 1960s with a focus on water resources, particularly flooding. He found himself working on a new geography of perception, the world inside people's minds (Johnston and Sidaway 2004). A major exponent of this geography with a behaviourist approach was Kates (1962: 1) who, in the introduction to his study on

74 Disaster management

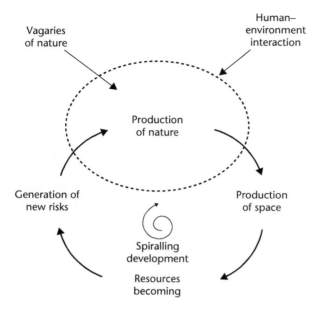

FIGURE 4.1 Human–environment interactions
Source: O'Brien *et al.* (2010).

floodplain management, wrote, 'The way men [*sic*] view the risks and opportunities of their uncertain environments plays a significant role in their decisions as to resource management.'

Kates built his understanding on the work of Simon (1957), whose approach to decision making in economics was to reject optimum economic behaviour of profit maximisation in favour of a bounded rationality, where people satisficed making 'bounded rational decisions' in pursuing economic strategies. A simple example is of people who choose to live in a coastal earthquake zone, such as San Francisco or Kobe, because the fault lines offer deep silt-free harbours for the economic development of marine transport while simultaneously harbouring levels of earthquake risk. There was, however, a significant critique of this approach. Notably, Smith and O'Keefe (1980, 1985) argued that geography displayed three major ways in dealing with people–environment relations in hazards research, which were the consequences of a positivist approach to the hazard question. McManus (2000) describes the critique as withering and lists the three major ways with which Smith and O'Keefe (1980) illustrate the poverty of the hazards research school where:

1. nature is separated from human activity;
2. nature is seen as natural and only hazardous when it intersects with human activity (Burton *et al.* 1993); and
3. humans are assumed to be absorbed by nature.

McManus (2000: 217) further notes, 'The first approach focuses attention on "natural causes" of disasters, rather than human vulnerability; the second is presented as a technocratic agenda to control nature, while the third is seen as Malthusian because it blames the victims.'

The reasons for this strong critique rest in earlier work by O'Keefe *et al.* (1976) highlighting the importance of understanding social vulnerability, which implied changing levels of risk in changing conditions of political economy. A parallel critique of the hazards paradigm emerged in the 1980s, sensitive to the globalising tendency of the paradigm and demanding a more progressive, people-focused approach to disaster planning (Hewitt 1983). Taken together, this people focus suggests there should be new ways of learning about disaster management, but little work has been attempted on learning about known unknowns, such as accelerated climate change and increased variability.

In the developed world, as the Cold War era began to draw towards its end, disaster management started to focus on an 'all-hazards' approach in response to a range of both natural and technological disasters. The all-hazards approach aims to reduce risk from civil, natural, technological, biological and instrumental disruptions. This approach incorporates response, recovery, mitigation and preparedness (McEntire *et al.* 2003). Figure 4.2 illustrates the disaster management cycle of the all-hazards approach.

In the disaster management cycle, the disaster management function is focused on preparedness for an event and then responding to that event. Recovery is usually the responsibility of other bodies, such as local government. Mitigation efforts are similarly the responsibility of other bodies, such as local government, but typically there will be liaison with the disaster management function about the type of mitigation measures, for example the scale and location of flood defences. This is a professionalised approach where different parts of the disaster management cycle are handled by specialist groups charged with distinct responsibilities. There have

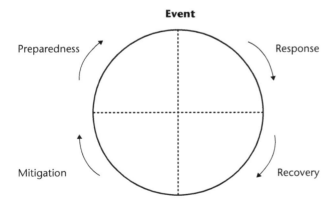

FIGURE 4.2 Disaster management cycle
Source: O'Brien *et al.* (2010).

been attempts to formulate this into a more comprehensive paradigm (McEntire et al. 2003).

There are, however, critiques of what has emerged as the dominant form of disaster management, particularly in the developed world, which is based on two misleading assumptions about people and communities.

First, other forms of social organisations, such as voluntary and community-based organisations, informal social groupings and families, are viewed as irrelevant or disruptive to disaster management because they are not controlled by the authorities. The second assumption is that disasters produce passive 'victims' who are overwhelmed by crises, premised on the view that the first responders are the disaster management professionals; however, the reality is that the first responders are actually the 'survivors'. There are many examples of survivors helping others in a disaster situation and sufficient research to refute such assumptions. Similarly, views that disaster-affected people panic or loot or simply serve their own interests are misleading. The conventional view is that 'victims' need to be told what to do and any aberrant behaviour needs to be controlled, even to the point of imposing martial law. There is considerable sociological research that contradicts these views (Quarantelli et al. 1972). Despite such research, there are still some media reports that seem to be unaware of new understandings of how people react in disastrous situations. This can lead to gross distortions, for example media reporting during Hurricane Katrina described a young African–American carrying a bag and beverages as 'looting a grocery store' and portrayed a white couple doing the same 'after finding bread and soda from a local grocery store' (Kaufman 2006). Contradictions such as this undermine the conventional view of disaster management, though this should be qualified by the view that for routine events, such as road traffic accidents, the current system works well. It is when the boundary between a small event and a disaster is crossed that rational thinking seems to disappear.

Internationally, thinking about disaster management has shifted from 'managing' disasters to prevention. This shift in thinking has been driven by a growing recognition that natural hazards have a significant impact on societies and a disproportionate impact on the developing world. Between 1991 and 2005, almost a million deaths and tens of billions of USD in damages resulted from the impact of natural hazards (Schipper and Pelling 2006).

The international response to this growing trend was the declaration that the 1990s would be known as the International Decade for Natural Disaster Reduction (IDNDR) by adopting resolution 44/236. A conceptual shift was signalled by a themed mid-term review of IDNDR, the Yokohama Strategy, in 1994, which requested practitioners and organisations to adopt a new paradigm that incorporated a greater emphasis on prevention rather than just reactive action. This initiated a move from disaster management to disaster risk management, often described as preparedness practice (Sperling and Szekely 2005).

In 2000, the UNISDR was established to build on the gaps and challenges identified by the Yokohama Strategy. The mission of UNISDR is to build disaster-resilient communities by promoting increased awareness of the importance of

DRR as an integral component of sustainable development, with the goal of reducing human, social, economic and environmental losses due to natural hazards and related technological and environmental disasters (UNISDR 2004). UNISDR places prevention at the heart of disaster management and clearly articulates the view that DRR should determine how we conceptualise disaster management. UNISDR (2009: 10) defines DRR as:

> The concept and practice of reducing disaster risks through systematic efforts to analyse and manage the causal factors of disasters, including through reduced exposure to hazards, lessened vulnerability of people and property, wise management of land and the environment, and improved preparedness for adverse events.

UNISDR clearly identifies the two components of risk: hazard and vulnerability. Furthermore, it clearly identifies the importance of the policy framework and acknowledges that adverse events will occur by advocating a focus on preparedness. Perhaps the most significant shift is the focus on people. There is acknowledgement that governments are the most appropriate body to promote and manage DRR, but this cannot be done without the participation of people and communities. Essentially, this marks a shift from a command-and-control approach to an approach that can be described as a partnership between government and people.

The mid-term review of UNISDR produced the *Hyogo Framework for Action 2005–2015: Building the Resilience of Nations and Communities to Disasters* that was adopted at the World Conference on Disaster Reduction. This framework sets out a plan of action for developing national platforms for DRR.

In summary, the concept of disaster management has shifted from a natural hazards and responsive focus to recognition that new hazards evolve along with new developments. The all-hazards approach attempts to address current and new hazards but retains an event focus. This is clearly seen by comparing the UNISDR definition of disaster risk management with that of disaster risk reduction given above. UNISDR (2009: 10) defines disaster risk management as:

> The systematic process of using administrative decisions, organisation, operational skills and capacities to implement strategies, policies and coping capacities of the society and communities to lessen the impacts of natural hazards and related environmental and technological disasters.

Conceptually this differs from DRR, which has its emphasis on prevention and preparedness and a clear focus on people.

Evolution of disaster management practice

Modern or conventional disaster management in the developed world after the Second World War was mainly focused on civil defence. The driver for this approach was the threat of nuclear attack during the Cold War era. As the threat of nuclear war subsided, greater attention was given to civil protection. This shift to a focus on civil protection should also be viewed against the profound changes during the twentieth century that shaped the risk landscape and societal attitudes to risk. The developed world saw great social and technological changes in the latter half of the twentieth century. Individualism rose with the information revolution and the emergence of a highly educated and increasingly mobile information society.

The corollary to this was the transfer of loyalties from institutions and structures to individual values or individualism in what Ulrich Beck terms reflexive modernity. Beck argues that the Industrial Revolution, the first modernity, saw many radical changes in everyday life, yet it was still based on traditional social structures, particularly family and gender. The latter half of the twentieth century, the second modernity, saw further changes, for example women entering the workplace, the shift from full-time to part-time employment, the erosion of lifetime job security in both blue-collar and white-collar occupations, and changing family and social structures, which began to modernise the foundations of the first modernity, making it reflexive (Beck 1992). This latter part of the twentieth century thus became an era that called into question both the role and legitimacy of institutions and structures. The increasing scepticism of officialdom is illustrated by the BSE (bovine spongiform encephalopathy) outbreak in the United Kingdom (see Box 4.1).

BOX 4.1

THE UNITED KINGDOM AND BSE

The first reported case of BSE in the United Kingdom occurred in 1986 and, despite warnings from the scientific community, the government failed to act or show concern about the possible transfer to humans. Tragically, transfer did occur in a new variant form of Creutzfeldt–Jakob disease (CJD), details of which were leaked to the press in 1996. The intervening period between the first case of BSE and the first case of CJD was punctuated by denials and reassurances by officials and politicians in response to growing public concern. The impact on the public was dramatic and trust in the UK Government effectively vanished. De Marchi and Ravetz (1999), in their analysis of the problems of risk and governance drawn from a number of case studies, characterise the situation in the following way:

> For the deeper problems of governance, the BSE case is the watershed. Whatever the ambivalences in the governance of risk that were shown at Seveso (and other cases like Chernobyl), it was only with BSE that it became elementary prudence to adopt the motto 'Don't believe it until it is [sic] been officially denied.'

(de Marchi and Ravetz 1999: 755)

It was this time of change that saw the emergence of the 'all-hazards' or comprehensive approach that incorporates response, recovery, mitigation and preparedness (McEntire et al. 2003; Figure 4.2). Acceptable levels of risk are embedded in norms and standards and are part of the governance structure that regulates societal activities. General acceptance by the public of the regulatory framework governing risk management is a necessary component of effective civil protection along with an emergency management function that meets societal expectations.

The all-hazards approach can be characterised as legally based, professionally staffed, well funded and organised. Training and exercising is aimed at improving the response function of emergency management professionals, evaluating communications systems, standardising procedures, testing and evaluating equipment and investigating stress on the responders. Alerting the public to potentially hazardous events is done through a variety of warning and informing systems. The aim is for a return to normality, i.e. to re-establish conditions as they were prior to the event (Perry and Peterson 1999; Schaafstal et al. 2001; Paton and Jackson 2002; Cassidy 2002; Perry and Lindell 2003). It is a pre-dominantly top-down technocratic model with an internal focus on organisational preparedness and institutional resilience. Table 4.1 summarises this approach.

The all-hazards approach to disaster management is intended to deal both with current and emerging risks. In Europe, and particularly the United Kingdom, the terrorist attacks of 11 September 2001 in the United States skewed the reform process, placing a greater emphasis on the terrorist threat and institutional resilience (O'Brien 2006). The United Kingdom is no stranger to terrorism. Irish Republican terrorism aimed, usually, at politically or economically sensitive targets in London and other parts of the United Kingdom has occurred over a number of years. But the attacks of 11 September added a new dimension, where the terrorist has no concern for their own life and is intent on causing as many fatalities as possible. This raised fears that terrorism of a new kind, organised, well financed and

TABLE 4.1 Technocratic model of disaster management

Dominant paradigm	Comment
Isolated event	Disasters usually regarded as unusual or unique events that can exceed coping capacity
Risk not normal	Risk is socially constructed and risk management aims to reduce risk to within proscribed levels realised through governance structures
Techno-legal	The legislative framework, regulatory system and the technologies used for risk reduction and disaster response
Centralised	Realised through a formal system such as a government department or state funded agency
Low accountability	Typically internalised
Post-event planning	Internal procedure for updating and validating plans based on lessons learned
Status quo restored	The overall aim – a return to normal

Source: O'Brien and O'Keefe (2010).

planned, ruthless and determined was about to be unleashed. Such fears were well founded. On 7 July 2005, London suffered bomb attacks to its transport system, demonstrating that no matter how well prepared, it appears almost impossible to prevent such atrocities, particularly if the terrorists themselves have no interest in surviving the attack. There is little doubt that the terrorist threat influenced preparedness in the United Kingdom. The recent London Olympic Games have seen unprecedented security measures and when the private contractor, G4S, failed to deliver sufficient trained security staff, additional UK military resources were deployed.

But it is the events leading up to the reform of the UK emergency response where it can be seen that the reforms were focused both on security and institutional resilience. After an emergency response, lessons learned from the event that can be incorporated into future responses become part of the post-planning activities. The objective of the emergency management cycle is to restore the status quo, for example after the floods in the United Kingdom in 2000, damaged homes were refurbished despite the fact that these areas are likely to flood again. Following a number of disruptive events (Table 4.2) prior to and after the turn of the millennium, it became clear that the UK system was in need of reform.

The process of wholesale reform began in 2001. It mapped out and implemented a legislative and capacity building programme under the banner of UK Resilience. A new legislative framework, the Civil Contingencies Act, was introduced in 2004. By and large, the reforms were welcomed (O'Brien and Read 2005). The Civil Contingencies Secretariat that led the UK reforms defined resilience as 'The ability at every level to detect, prevent, prepare for and if necessary handle and recover from disruptive challenges.' (Great Britain Cabinet Office Civil Contingencies Secretariat 2004: 1)

The UK reforms, though needed, retained a top-down command-and-control structure and it is evident that attempts at resilience building were aimed at institutions.

The UK Civil Contingencies Act requires that UK emergency services have continuity plans that apply to all of their functions, not just emergency response functions. Additionally, local authorities are required to provide business continuity management (BCM) advice to businesses and voluntary organisations. This is an

TABLE 4.2 Timeline of events

Year	Type of event
1987	Fire disaster at King's Cross railway station
1987	Ferry disaster at Zeebrugge, Belgium
1988	Clapham rail crash
1989	Hillsborough football stadium disaster
2000	Y2K Millennium Bug: fears of mass computer crashes
2000	Floods: recurrent heavy rainfall led to several severe flooding events
2000	Fuel blockade: disruption to transport of fuel deliveries
2001	FMD: foot and mouth outbreak

interesting departure from previous practice because it suggests that this new duty for local authorities is meant to cover not only responses to disruptions, such as a power failure or fire, but also responses relevant to the new realities of terrorist attacks (Savage 2002; Seifert 2002). These policies clearly reinforce institutional and organisational response at the local level.

But it was the changes at regional and national levels that were significant. While Part 1 of the Civil Contingencies Act strengthens the local level, Part 2 effectively provides a clean sweep. It repeals legislation covering civil protection emanating from 1948 and the 1920 Emergency Powers Act, deemed inadequate to meet modern threats such as chemical, biological, radiological and nuclear (CBRN) terrorist attacks. The Civil Contingencies Act allows the use of emergency powers either nationally or on the basis of the English regions and devolved administrations of Scotland, Wales and Northern Ireland and requires the appointment of a regional nominated coordinator who will act as the focal point if the emergency powers are used. The scope of the emergency powers set out in Section 22 of the Act is fairly broad and includes the use of military forces, the confiscation of property and the prohibition of movement or forced movement in the event of an emergency. What is clear from the guidance to the Civil Contingencies Act is that emergency regulations made under the Act cannot include the full range of powers set out in Section 22 and are open to parliamentary scrutiny and challenge.

The Civil Contingencies Act establishes Regional Resilience Forums (RRF) staffed by officials that provide a single line of communication and information sharing from the local level through to the Civil Contingencies Secretariat at the centre, leading to the possibility of subversion into a command-and-control structure. Militarisation during a crisis is a disturbing but not unrealistic prospect and was seen with refugees being herded into the Superdome by armed national guardsmen during Hurricane Katrina (Wisner 2005). Whether or not we will see the return to a more authoritarian approach as a result of the 7 July attacks in London is not clear but it is a real concern that the changes towards more civilian approaches following the end of the Cold War might be abandoned in times of crisis (Alexander 2002).

The reforms are claimed to have improved UK resilience but the Capabilities Programme, the core framework for building resilience throughout the United Kingdom, shows what the government's real priorities were. While resilience should be a holistic approach, the government approach has been to narrowly focus on institutional resilience. The programme included seventeen 'work-streams' that fall into three groups:

1. Three work-streams are essentially structural, dealing respectively with the central (national), regional and local response capabilities.
2. Five work-streams are concerned with the maintenance of essential services (food, water, fuel and transport, health and financial services).
3. Nine functional work-streams dealing respectively with the assessment of risks and consequences; CBRN resilience; human infectious diseases; infectious

diseases of plants and animals; mass casualties; mass fatalities; mass evacuation; site clearance; and warning and informing the public.

Although each work-stream is the responsibility of a lead government department, the Civil Contingencies Secretariat sits at the centre and effectively exercises hegemony – a centralised approach with command-and-control overtones. There is also considerable resonance between the nine functional workstreams and the modalities of a terrorist threat, either a suicide bomb attack or CBRN terrorism. Mass casualties, fatalities and evacuations are the modalities of response to terrorist attacks and the capacity for the site clearance part of the longer-term recovery and return to normality. There is little doubt that the UK government is focused on terrorism and institutional resilience (O'Brien 2006).

Since the Civil Contingencies Act was enacted there has been some attempt to build social resilience. The Civil Contingencies Secretariat did start a consultation process; however, since the change of government in 2010 little progress has been made.

Post-11 September, the emphasis and responsibility for disaster management in the United States shifted as the Federal Emergency Management Agency (FEMA), established in 1979 to bring together a number of agencies responsible for civil protection, was incorporated into the newly formed Department of Homeland Security (DHS). The focus of the new department on security and changes in budget priorities weakened the effectiveness of FEMA's all-hazards programmes (Berginnis 2005; Elliston 2004). The inadequate response and performance of the FEMA and DHS bureaucracy during and after hurricane Katrina illustrates the problems caused by the new structures (Dynes and Rodriguez 2006).

The contextual changes to disaster management in the United States and Europe can be characterised as having moved from the militaristic command-and-control approach of the bi-polar era towards a more participatory and collaborative system and back to the militaristic style post-11 September (Alexander 2002). In Europe, and particularly the United Kingdom, the 11 September terrorist attacks have reinforced an institutional and security focus that has been further strengthened in reaction to the London and Madrid bomb attacks (O'Brien 2006). Although the overall principles of response, recovery, mitigation and preparedness underpinning the disaster management cycle have not changed, the emphasis is more focused on institutional preparedness and security as opposed to public preparedness.

It can be concluded that the current practice of disaster management, particularly in Europe and North America, requires a shift in emphasis if it is to focus on DRR, but there is danger of oversimplification in making too much of a generalisation about disaster practice. The practice of disaster management is conditioned by the nature of hazards most usually experienced, for example countries such as Bangladesh experience cyclones whilst others such as Japan are susceptible to earthquakes. Thus, the structure of the disaster management function evolved within the national context, for example is it centralised or decentralised, is partnership working encouraged or discouraged and what is the nature of linkages to broader civil society? As the disaster management function will have evolved in particular circumstances, practices and structures are likely to be highly varied.

Despite these variations, there are two types of approach to risk reduction in disaster management practice that are common to all disaster management functions (although the extent to which these are practised will vary) – namely structural and non-structural measures. UNISDR defines these as:

> **Structural measures:** Any physical construction to reduce or avoid possible impacts of hazards, or application of engineering techniques to achieve hazard-resistance and resilience in structures or systems.
> **Non-structural measures:** Any measure not involving physical construction that uses knowledge, practice or agreement to reduce risks and impacts, in particular through policies and laws, public awareness raising, training and education.
> (UNISDR 2009: 28)

Structural measures refer to any physical construction that will reduce or avoid, as far as possible, the impacts of hazards. These include engineering measures and construction of hazard-resistant and protective structures and infrastructure and can include dams, flood levies, ocean wave barriers, earthquake-resistant construction, evacuation shelters, etc.

Non-structural measures refer to policies, awareness, knowledge development, public commitment, and methods and operating practices, including participatory mechanisms and the provision of information, which can reduce risk and related impacts. Common non-structural measures include building codes, land-use planning laws and their enforcement, research and assessment, information resources and public awareness programmes.

Although structural interventions can achieve success in reducing disaster impacts, they can also fail due to lack of maintenance, age or due to extreme events that exceed the engineering design level (Doyle *et al.* 2008; Galloway Jr. 2007; Galloway *et al.* 2009). Japan has the most heavily defended coastline in the world; however, the scale of the tsunami on 11 March 2011 overwhelmed defences developed to standards that did not anticipate an event of such magnitude (Asian Disaster Reduction Center (ADRC) and International Recovery Platform (IRP) 2011). On the other hand, in Oirase on Japan's northeast coast, a sea wall proved decisive in protecting citizens. This illustrates the difficulties with structural measures. Most structural measures have a specific design life at the time of construction and thus can be viewed more as short-term solutions with short-term benefits, which may or may not be sustainable in the longer term or under changing conditions, including climate change. Furthermore, technical considerations should not preclude local social, cultural and environmental considerations (Opperman *et al.* 2009; World Meteorological Organisation (WMO) 2003). Implementing structural measures from planning through to implementation that involve participatory approaches

84 Disaster management

with local residents who are proactively involved often leads to increased local ownership and more sustainable outcomes. It is becoming clearer that the measures themselves, particularly structural measures, are insufficient without a shift in thinking towards greater preparedness. This leads to an entirely different approach: community-based disaster management.

Community-based disaster management

The concept of community-based disaster management (CBDM), sometimes known as community-based disaster risk management (the term CBDM will be used throughout this book), emerged in South East Asian countries. In Bangladesh, the cyclones in 1970 and 1991 resulted in the deaths of 500,000 and 138,000, respectively. Although both events had catastrophic impacts on Bangladesh, it is important to note that following the 1970 disaster the government and other agencies began to implement the Bangladesh Cyclone Preparedness Programme, a bottom-up programme aimed at communities reducing their vulnerabilities and enhancing resilience. The national government worked in partnership with other agencies to develop a community-based approach to disaster management. This represented a different approach and a determination to learn from experience. An example of measures implemented is cyclone shelters. In the 1991 storm, fatality rates were 3.4 per cent in areas with access to cyclone shelters compared to 40 per cent in areas without access to shelters. Because of improved preparedness during another strong storm in 1994, three-quarters of a million people were safely evacuated and only 127 died (Schultz et al. 2005; Akhand 2003). It is worth contrasting this with Hurricane Katrina, which hit Louisiana in 2005 and was the costliest hurricane in the history of the United States, both in terms of lives lost (over 1800) and economic damages (some USD 81 billion). Hurricane Sandy in 2012, dubbed a Superstorm, caused at least 285 fatalities (over 70 in the US) and caused USD 75 billion worth of damage. Hurricanes are not unusual events in the United States. Florida, for example, experiences hurricanes regularly but has established building codes that are designed to minimise damage and has well-rehearsed evacuation procedures. It is noteworthy that a poor country, such as Bangladesh, can learn from experience, whereas in a sophisticated country, such as the United States, one state, Louisiana, in close proximity to another, Florida, could not learn from its experience.

The Hyogo Framework for Action 2005–2015 prioritises DRR at local and national level and the establishment of strong and functional institutions to manage disasters (UNISDR 2005). As a vehicle, CBDM offers a way of engaging with communities and making them more self-reliant. This is important because disasters are everyone's business and both the planning and preparedness functions should be part of community development. In that sense, CBDM, with the active participation of vulnerable communities, can help to identify local hazards and devise locally appropriate strategies and development activities to reduce disaster losses. Community participation in the development and implementation of these plans ensures ownership, which contributes to their sustainability. Using holistic approaches that incorporate the needs of local communities can provide the impetus for the development of locally owned, community-based, multi-stakeholder disaster management plans that are integrated with the periodic development plans of those areas within which these vulnerable communities reside. These plans should enable communities to prevent, reduce and effectively respond to stresses, shocks and potentially disastrous events. Indeed, the implementation of such plans is an essential component of poverty reduction and sustainable development.

In some ways CBDM is similar to conventional disaster management practice, for example the approach is premised on risk assessment. However, major differences emerge at this point. Risk assessments in CBDM are carried out in a participatory way. Communities assess their own hazards, risks and vulnerabilities. Various tools and methodologies are available for participatory risk assessment and they often need to be adapted to the local conditions. This is very different to the professional approach adopted in developed countries.

Community-based risk management has traditionally dealt with variability in weather conditions but long-term climate change and increasing variability will require more proactive behaviour at the community level. This will be challenging and will require a shift in the way local government and NGOs interact with local communities; a shift from reactive and often non-transparent modes to proactive approaches that are aimed at building community resilience. The barriers to resilience building broadly centre around two areas. First, the long-term nature of capacity building. Second, the cyclical nature of political election cycles, which is often focused on the short term. For example, although the consequences of a hurricane impacting New Orleans were understood in terms of planning and response, this information was ignored at all levels of government, including the local level (Kates et al. 2006). A window of opportunity does open after the occurrence of an event, particularly after extreme events. Whether or not recent events, such as the floods in Pakistan and droughts in the wheat-growing areas of China, will provide sufficient impetus is debatable. The lack of both a positive outcome at Durban and a sense of political urgency does not bode well.

Humanitarianism

DRR in developing countries is closely tied to the delivery of humanitarian aid. This is both an opportunity and a threat. In the first instance, any country hit by a natural disaster has responsibility to be the first provider of rescue and relief activities. The UN Humanitarian Resolution 46/182 of 1991 states:

> Each state has the responsibility first and foremost to take care of the victims of natural disasters and emergencies occurring on its territory. Hence, the affected state has the primary role in the initiation, organisation, coordination and implementation of humanitarian assistance within its territory.
>
> (United Nations 1991: Annex I, p. 4)

Fine words, but not necessarily close to the reality of humanitarian intervention. A little history might provide a deeper understanding of the problem. The notion of humanitarianism emerged in the nineteenth century when the French philosophers Leroux and Comte tried to define and structure, respectively, a humanitarian religion (Davies 2012). Core to this drive was to live life for others, transferring self-sacrifice from God to people. Harrison (1879), the translator of Comte, suggested that the religion of humanity went beyond merely improving the social conditions of people but, once Divine purpose is abandoned, ways can be sought to prevent natural disasters, which can be seen as preventable, 'The famines, the diseases and the revolutions which afflict mankind are no longer the judgements of God. They are the inevitable consequences of known and preventable conditions.'

It is notable that famine was seen as a human construction, that diseases were preventable and that revolutions occurred because there was insufficient attention to the political, social and economic conditions of the population. The humanitarian movement sought to improve conditions for workers, advance education for the poor and build cooperative communities (Charlton 1906).

At an international level, there was the emergence of the Red Cross, championed by Henry Dunant. It eventually functioned at three levels: the International Confederation of the Red Cross, which is a guardian of humanitarianism; the International Federation of Red Cross and Red Crescent Societies (IFRC); and national Red Cross and Red Crescent societies. It has principles for the delivery of humanitarian assistance, namely humanity, impartiality, neutrality and independence that are really operational requirements rather than statements of rights.

In parallel with these developments, especially during and after the Second World War, the UN was given the responsibility for leading humanitarian action (a crude periodisation of humanitarianism would establish four periods: pre-Red Cross, Red Cross, post-1945 and, finally, post-1989 and the ending of the cold war). Essentially, it was a re-think after the failure of the League of Nations to address the problem of German refugees in the run-up to the Second World War. The core issue was not the response to disaster but the problem of refugees, essentially a disagreement between the United States and the USSR. When the United Nations High Commissioner for Refugees (UNHCR) office was established in 1951, UNHCR was mandated to monitor UN member states' compliance with the 1951 Convention on the Status of Refugees. Over time, UNHCR's statutory competence increased. The image of individual refugees struggling across the Berlin Wall led to a flood of over 20 million refugees and 40 million internally displaced people after the Berlin Wall was broken down, a metaphor for the collapse of the Soviet Union.

During the Cold War, much of the disaster relief was focused on Africa. The Biafran War, 1967–1970, saw agencies, including international non-governmental organisations (INGOs) such as Oxfam, increasingly politicised in the delivery of humanitarian assistance. This was the first of the televised disasters, to be followed by the African droughts of 1973–1976 and 1984–1986. The landscape was studded with famine camps. The fall of the Soviet Union ushered in an era of complex emergencies, local wars, which were addressed with the same humanitarian machinery of the UN family, the Red Cross/Red Crescent movement and the INGOs. Médicins sans Frontiéres (MSF)/Doctors without Borders was established to speak out or bear witness, essentially, against nation states that generated vulnerability. For the rest, an oligopoly of INGOs emerged, frequently Christian, to push a Western model of aid onto a largely Muslim world where the conflicts were often generated by Western intervention. Somalia, Kosovo, Iraq and Afghanistan are the obvious ones, but intervention in Darfur, Sudan, was underpinned by a sense of Christian Western imperative. Tony Blair, speaking in the House of Commons before the Kosovo War started, justified that war would 'avert what otherwise would be a humanitarian disaster'. The military was linked to humanitarian intervention, humanitarian intervention based on humanity, impartiality, neutrality and independence was an arm of Western foreign policy.

The UN, as lead agency, has an effective division of labour that sees the Office for the Coordination of Humanitarian Assistance (OCHA) in charge of the policy and planning framework; the World Food Programme (WFP) responsible for emergency food delivery and logistics; the UNHCR responsible for shelter; the United Nations Children Fund (UNICEF) responsible for nutrition and water, sanitation and hygiene; and the Food and Agricultural Organisation (FAO) responsible for emergency agriculture, which, if successful, should mark the end of the emergency and the diminution of the role of the WFP. Medical interventions are largely left to INGOs and the Red Cross.

International government response reached a peak of USD 13 billion in 2010 from a starting point of some USD 0.5 billion with the collapse of the Soviet Union in 1989 (Global Humanitarian Assistance (GHA) 2012a). No other public sector expenditure matches that growth, with the exception of military expenditure. The gap in unmet financing (appeal total against actual response) continues to widen. WFP (food) received some 80 per cent of its appeal, UNHCR (shelter) 60 per cent and UNICEF (water and sanitation) less than 40 per cent between 2004 and 2009.

A number of countries, especially in Latin America and Asia, are developing their own systems, including the military, to respond to disasters and are seeking different relationships with the humanitarian community. Coupled to this is a sense of anger and frustration at humanitarian agencies' lack of deference to national authority and sovereignty and tensions between cultures (Harvey 2009). In turn, this is exacerbated by humanitarian operations on the ground that appear uncoordinated.

In one sense, a supply-side sense, humanitarian aid is very coordinated. Two donors, the United States and the EU, each account for one-third of global government aid through US Agency for International Development (USAID) and European Community Humanitarian Office (ECHO), respectively. The European Commission has championed the linking of disaster to development all the way to the European Parliament. This linking, strongly supported by INGOs, sees the poverty agenda at the centre of the process and connects to a human rights agenda. It also significantly distorts the humanitarian imperative, making it seem more Western policy in disguise.

Part of the debate has been to accommodate DRR, both into disaster response and development planning. Less than 1 per cent of aid goes to DRR activities (Global Humanitarian Assistance (GHA) 2012b). DRR, and by implication climate change and broader environmental issues, are not aid priorities despite it being noted at the recent Fourth High Level Forum on Aid Effectiveness (Busan, Korea) that donor governments '... must ensure that development strategies and programmes prioritise the building of resilience amongst people and societies at risk from shocks, especially in highly vulnerable settings' (Busan Partnership for Effective Development Cooperation 2011).

The use of resilience to bridge humanitarian assistance, DRR, good governance and development goals in one framework was highlighted through the recommendation of the UK government's Humanitarian Emergency Response Review, but to date the focus has been on natural disasters (Department for International Development (DFID) 2011a). DFID's definition of resilience is:

> Disaster resilience is the ability of countries, communities and households to manage change, by maintaining or transforming living standards in the face of shocks or stresses – such as earthquakes, drought or violent conflict – without compromising their long-term prospects.
>
> (DFID 2011b)

Transforming living standards would be a major step forward, but with neither the humanitarian nor the development community identifying with the problem and providing financial support, the message seems to be that developing countries do it themselves.

Hazards and risk, vulnerability and resilience

The terms hazard, risk and vulnerability are the most common associated with disaster management. The term resilience has entered the disaster management lexicon more recently and conceptually recognises the capacity of people and societies to bounce back or recover from disruptive events. This is discussed in more detail later in the book.

Hazards are regarded as the physical conditions, either naturally occurring or human made, which present a threat to life or property (Tobin and Montz 1997). More recently, the distinction between natural and human made has become somewhat blurred. As discussed earlier and illustrated in Figure 4.1, human–environment interactions modify the environment, for example the clear felling of trees to produce land for agriculture. Providing that the interaction is at a scale that is within the regenerative capacity of the area then that should not cause long-term problems. The interaction will, however, produce new kinds of hazards. Taking the conversion of land for agriculture as an example, there is a possibility that drainage could be affected in ways that could lead to flooding. But bigger problems can occur when we overstep ecological limits – and that can have catastrophic consequences.

Jared Diamond in his book *Collapse: How Societies Choose to Fail or Succeed* lists eight environmental factors that have contributed to societal collapse (Diamond 2005). He cites the Easter Island collapse as being solely due to environmental damage. In the case of the Anasazi of south-western North America and the Mayans of Central America, environmental factors of degradation and climate change contributed significantly to collapse. Diamond goes on to list four environmental factors that are influencing the modern world: climate change, build-up of toxins in the environment, energy shortages and the full human utilisation of the Earth's photosynthetic capacity. As shown in Figure 4.3, the vagaries of nature are the cause of natural hazards; however, human–environment interactions are of such a scale that we are producing both new kinds of vagaries and more extreme events. We have entered an era where we are producing the hazards we face. There are many other types of hazards we produce, such as the global credit crunch, which are not related to the vagaries of nature but are clearly a product of the vagaries of man.

Our reckless disregard is influencing and modifying the environment on which we depend. Disruption of the ecosystem is not new and throughout its history

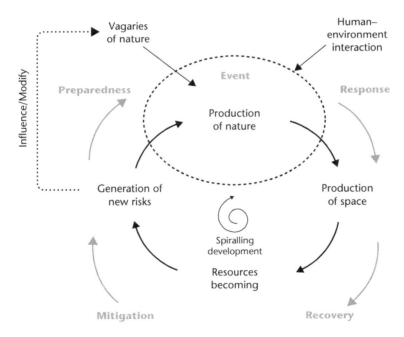

FIGURE 4.3 Conventional disaster management within human–environment interactions

Source: O'Brien *et al.* (2010).

the Earth has experienced disruptive events and mass extinctions from which it has recovered. Lovelock postulates in his Gaia Theory that the ecosystem acts to preserve the optimum conditions for life on the planet and has the capacity to recover from disruptions but, as Lovelock points out, life on the planet does mean just human life (Lovelock 2006). A collapse driven by our actions could lead to wholesale societal collapse. It may not mean mass extinction of people but it could lead to a situation where a few ragged bands of survivors seek somewhere to eke out a living. Not a tempting prospect.

This is somewhat at odds with the concept of disaster management that seeks to protect life and property. All governments have a tacit agreement with their citizens that they will protect them from adversity. This approach has become embodied in the 'all-hazards' approach. Climate change presents a new form of hazard for which this model is not entirely appropriate. The anthropogenic production of GHGs by, primarily, the industrialised countries is generating a new family of produced unknowns. Though known in general terms, the 'what' and the 'when' cannot be predicted with any accuracy. It is generally accepted that anthropogenic activities are driving changes in the climate system (IPCC 2007a). This can be described as a 'wicked problem' because there is little opportunity to learn by trial and error or any exit point from the problem (Ritchey 2007; see also Chapter 3).

One problem is that the nature of climate hazards is different from the nature of hazards that are envisaged for disaster management. Table 4.3 illustrates these differences.

As can be seen in Table 4.3, the nature of the hazards produced by climate change is very different to the discrete events normally envisaged by disaster management. Essentially, disaster management is predicated on responding to a discrete event driven by natural variability, such as seasonal events that are reasonably certain, for example the Monsoon season in Asia or cold weather in northern latitudes. The changing nature of climate-driven events means that the risk management systems underpinning disaster management are weakened. We will need to do things differently.

Risk is the combination of the probability of a hazard event occurring and the anticipated magnitude of loss associated with it (in terms of mortality, well-being, economic, environmental, resources) (Tobin and Montz 1997; Wisner *et al.* 2004).

TABLE 4.3 Contrasting characteristics of the traditional view of natural hazards and the new perspective presented by climate change

Hazard characteristic	*Natural hazards*	*Climate change*
Temporal	Discrete events	Long-term and continuous
Dynamics	Stationary	Non-stationary
Spatial scope	Regional	Global but heterogeneous
Uncertainty	Low to medium	Medium to very high
Attribution	Natural variability	Natural and anthropogenic

Source: Adapted from Fussel (2005).

TABLE 4.4 Definitions of vulnerability

Author/date	Definition
Timmerman, 1981	Vulnerability is the degree to which a system acts adversely to the occurrence of a hazardous event. The degree and quality of the adverse reaction are conditioned by a system's resilience (a measure of the system's capacity to absorb and recover from the event).
Susman *et al.*, 1983	Vulnerability is the degree to which different classes of society are differentially at risk.
Bogard, 1989	Vulnerability is operationally defined as the inability to take effective measures to insure against losses. When applied to individuals, vulnerability is a consequence of the impossibility or improbability of effective mitigation and is a function of our ability to detect hazards.
Mitchell, 1989	Vulnerability is the potential for loss.
Liverman, 1990	Distinguishes between vulnerability as a biophysical condition and vulnerability as defined by political, social and economic conditions of society. She argues for vulnerability in geographic space (where vulnerable people and places are located) and vulnerability in social space (who in that place is vulnerable).
Downing, 1991	Vulnerability has three connotations: it refers to a consequence (e.g. famine) rather than a cause (e.g. drought); it implies an adverse consequence (e.g., maize yields are sensitive to drought; households are vulnerable to hunger); and it is a relative term that differentiates among socioeconomic groups or regions, rather than an absolute measure or deprivation.
UNDRO, 1991	Vulnerability is the degree of the loss to a given element or set of elements at risk resulting from the occurrence of a natural phenomenon of a given magnitude and expressed on a scale from 0 (no damage) to 1 (total loss). In lay terms, it means the degree to which individual, family, community, class or region is at risk from suffering a sudden and serious misfortune following an extreme natural event.
Dow, 1992	Vulnerability is the differential capacity of groups and individuals to deal with hazards, based on their positions within physical and social worlds.
Smith, 1992	Human sensitivity to environmental hazards represents a combination of physical exposure and human vulnerability plus or minus the breadth of social and economic tolerance available at the same site.
Alexander, 1993	Human vulnerability is a function of the costs and benefits of inhabiting areas at risk from natural disaster.
Cutter, 1993	Vulnerability is the likelihood that an individual or group will be exposed to and adversely affected by a hazard. It is the interaction of the hazard of place (risk and mitigation) with the social profile of communities.
Watts and Bohle, 1993	Vulnerability is defined in terms of exposure, capacity and potentiality. Accordingly, the prescriptive and normative response to vulnerability is to reduce exposure, enhance coping capacity, strengthen recovery potential and bolster damage control (i.e. minimise destructive consequences) via private and public means.

Continued

TABLE 4.4 Continued

Author/date	Definition
Blaikie et al., 1994	By vulnerability we mean the characteristics of a person or a group in terms of their capacity to anticipate, cope with, resist and recover from the impact of a natural hazard. It involves a combination of factors that determine the degree to which someone's life and livelihood are put at risk by a discrete and identifiable event in nature or in society.
Green et al., 1994	Vulnerability to flood disruption is a product of dependence (the degree to which an activity requires a particular good as an input to function normally), transferability (the ability of an activity to respond to a disruptive threat by overcoming dependence either by deferring the activity in time, or by relocation, or by using substitutes), and susceptibility (the probability and extent to which the physical presence of flood water will affect inputs or outputs of an activity).
Bohle et al., 1994	Vulnerability is best defined as an aggregate measure of human welfare that integrates environmental, social, economic and political exposure to a range of potential harmful perturbations. Vulnerability is a multilayered and multidimensional social space defined by the determinate, political, economic and institutional capabilities of people in specific places at specific times.
Comfort et al., 1999	Vulnerability are those circumstances that place people at risk while reducing their means of response or denying them available protection.
Weichselgartner and Bertens, 2000	By vulnerability we mean the condition of a given area with respect to hazard, exposure, preparedness, prevention, and response characteristics to cope with specific natural hazards. It is a measure of capability of this set of elements to withstand events of a certain physical character.

Source: Adapted from Weichselgartner (2001).

With climate change, both the probability and magnitude and associated losses are increasing, but there is more uncertainty because of produced unknowns. We do know that impacts will vary across regions, which implies that risk management strategies will have to be tailored to regional and local needs.

The key determinants of disaster risk are exposure and vulnerability. Exposure is defined as:

> The presence of people; livelihoods; environmental services and resources; infrastructure; or economic, social, or cultural assets in places that could be adversely affected.
>
> (IPCC 2012a)

Vulnerability is defined as:

> The propensity or predisposition to be adversely affected.
>
> (IPCC 2012a)

However, because of vulnerability's multifaceted nature there is no commonly agreed definition of the concept (Bohle 2002). The current literature encompasses more than 25 different definitions, concepts and methods to systematise vulnerability (Birkmann 2006). Some of the definitions are given in Table 4.4.

In summary

Although risk management is an established discipline, the practice is based on what is best described as actuarial records, which are data collected over time, such as accident and damage statistics. There are other sources of data, such as computer modelling, but the essence is that we have little experience or knowledge of life in a slowly changing environment. IPCC relies on sophisticated computer modelling to develop scenarios of what might happen for different emission trends. From its Fourth Assessment Report, the computer models show little difference in global average temperature increase up to 2030 (IPCC 2007a). After this period there is divergence between the scenarios. From a political perspective, 2030 seems a long way off and the lack of difference between scenarios up to 2030 does not give much incentive to take radical and innovative decisions. Politics has its focus on the immediate future and the ongoing financial crisis has transfixed decision makers.

There is a great deal of uncertainty about climate change. In earlier parts of this book, we discussed that there is a growing number of researchers who believe that IPPC estimates are too conservative and the likelihood of being able to maintain the average global temperature rise below 2 degrees Celsius are slim. More recent evidence suggests that methane, a very potent GHG, is being released in the Arctic regions. Scientists have identified some 150,000 methane seeps in Alaska and Greenland. They also point out that the Arctic is the fastest warming region on the planet and have detected seepage along the permafrost thaw boundaries and in the moraines and fjords of retreating glaciers (Black 2012). This again points to the uncertainties in climate science, and methane seepage could act as a 'tipping point' where the additional warming caused by methane produces more methane seepage, which further accelerates the warming trend. Such news should act as a rallying call to political leaders, but there appears to be an institutional disconnect between decision makers and the scientific community.

In summary, it is clear that disaster management practice has evolved and there is an international recognition that we must shift from a reactive to a proactive stance with people very much the focus and the partner in DRR. There is still a long way to go. Regrettably, the culprit for the lack of progress in shifting the focus of disaster management in many instances is the government. Governments do not like change, but the threat of climate change may force them to change.

5
ADAPTATION

> There is an art to science, and a science in art; the two are not enemies, but different aspects of the whole.
>
> If knowledge can create problems, it is not through ignorance that we can solve them.
>
> (Isaac Asimov (1920–1992))

Introduction

Both authors are very concerned about the global environmental challenges we face. Evidence suggests that things will get worse and strategies for climate adaptation are needed urgently. Other parts of this book have discussed trends such as globalisation, growing population and the failure of the neo-liberal economic model as contributors to the degraded state of the Earth. Although both authors have been involved in development, disaster management and climate change for a number of years, it was work conducted on the Netherlands Climate Assistance Programme (NCAP[1]) that provided the impetus for this book. From 2003 to 2007, NCAP worked with 14 developing countries to develop climate adaptation policies. The findings, discussed next, are the basis for this book (O'Brien *et al.* 2011).

The adaptation continuum

Resilience must address vulnerability. The key to lowering vulnerability is poverty alleviation.

Poverty alleviation is an integral part of development if individual and community well-being is to increase. Poverty is not just defined by crude measures of daily personal income, although global comparisons using such data are illuminating. Poverty is defined locally in Bolivia as 'having no friends'; in Ghana as having no 'ancestral connections'; in India poverty is about 'not being free from the vices of greed'. Poverty is not just a state but also a process of social relationships, which are constructed in order to survive. Traditional extended family and social network obligations and entitlements are built and eroded as everyone is drawn into the global market place.

In poverty, people face the 'vagaries of nature', and more so as extreme weather events increase as an outcome of climate change. Development seeks, among other things, a natural environment that is more uneventful. Ironically, it does this most effectively when it enhances both natural and social systems' diversity. Diversity is essential to building resilience; social diversity is the key to building community resilience. Resilience success gives a positive feedback to questions of impacts and vulnerability, coming back to the initial starting point of climate adaptation. We call this process, shown in Figure 5.1, the 'adaptation continuum', which is discussed in detail later in this chapter.

The first stage in the adaptation continuum is:

- impacts to vulnerability – from 'what if' studies of future climate scenario impacts to understanding present vulnerability; who is exposed to what, where and when.

Three further key stages in the adaptation continuum are:

- vulnerability to adaptation – shifting the focus from current exposure to processes of adjusting to current, expected and potential resources and risks;
- adaptation and development – embedding climate adaptation in development planning, recognising the complexity of multiple stresses; and
- development to resilience – viewing adaptive management as a process of ensuring the resilience of development pathways.

This proposition of an adaptation continuum is consistent with the principles of climate adaptation as a socio-institutional process rather than the outcome of scenarios of vulnerability with some anticipated adjustment. This process-based understanding is in stark contrast to the static view of adaptive capacity as a snapshot of status indicators, such as GDP per capita.

From vulnerability to adaptation

Much of the literature recognises that reduced crop yields, expanded zones of vector-borne diseases, sea level rise and other effects of climate change will affect the most vulnerable because the poor have the least capacity to adapt. The key question regarding adaptation scope or coverage has to do with what matters to those who are most vulnerable to climate change impacts. Developing a framework to better understand the coverage needed in the shift from vulnerability to adaptation concerns metrics of impact that focus on the direct effects on community

FIGURE 5.1 The adaptation continuum

or household assets. In other words, it involves considering what is at risk and what/how much is potentially lost. This is particularly true when thinking about financial, institutional and social assets currently enjoyed but increasingly at threat without appropriate adaptation interventions. What matters in terms of developing adaptation measures are the livelihood assets that characterise the 'sustainable livelihoods approach': financial, social, physical, personal and human. Adaptation solutions need to relate to these assets. Consequently, the scope or coverage for this transition segment of the adaptation continuum (i.e. from vulnerability to adaptation) implies careful consideration of the specific assets at risk.

In adaptation, scale is a paramount issue. At what scale – local, regional, national or international – adaptation processes should take place is a challenging issue. Moving the debate from vulnerability assessments to adaptation requires the application of different tools and methodologies that allow for the integration of information and concerns. What has been learned is that there is no single methodology because there is a range of problems that need to be tackled, but all methodologies must be participatory and owned by those who are researched. The NCAP projects used approaches that focused on social learning/stakeholder consultation and participatory rural appraisal.

This is expressed in the way the projects have used different sets of tools and methodologies to address country-specific vulnerabilities. Bolivia, for example, adopted both a participatory and integrated approach to assess vulnerability and develop adaptation strategies for two sectors: human health and production systems. The country combined different methodologies and tools that considered biogeophysical factors, local perceptions, organisational settings and traditional knowledge to build adaptive capacity and facilitate social learning for adaptation.

Perhaps the most critical element in the vulnerability to adaptation stage is the integration of the obtained outputs in the political or policy dynamic. It makes little difference to apply methods and tools to identify the most suitable adaptation initiatives or to develop innovative communication protocols to transfer the results to decision-makers if these activities do not result in concrete outputs supported by budget lines, new legislation and/or leverage of new financial sources. Affecting the political and policy dimensions must be the ultimate test of efficacy of the vulnerability to adaptation process. Integrating the outputs in political and policy dynamics requires engaging politicians through lobbying, mobilising public support through information campaigns and steering the attention of powerful ministries (i.e. finance ministry and planning ministry) towards these outputs.

Adaptation and development

Adaptation actions based on vulnerability assessments are entrained in development planning; similarly development planning demands specific information on climate adaptation strategies and measures. Trying to move the adaptation agenda into development planning requires the adoption of a new perspective. In many senses, the adaptation–development perspective is somewhat parallel to successful pre-disaster planning, but pre-disaster planning itself has rarely managed to engage with the development agenda. While there has been continuous discussion of the relief through development continuum, the debate has treated pre-disaster planning and development as separate entities instead of focusing on their synergies and potential contribution to effective planning. A successful adaptation–development agenda could substantially reduce the cost of emergency disaster assistance. In the event of simultaneous disasters, increasingly likely as climate change accelerates, the increased demand on national and international disaster relief bodies could overwhelm local coping capacity. Self-reliance, realised through effective pre-disaster and adaptation planning as an integral part of development and aimed at capacity building for the most vulnerable, is a more effective means of DRR. This approach builds resilience to respond to and recover from climate change impacts and is more effective than a reactive post-event approach.

Strategies for adaptation to climate change combine relief, reconstruction and rehabilitation, seeking to promote sustainable conditions and self-reliance. Integrating adaptation into development planning broadens the metric of impact beyond direct effects (e.g. economic damage and lives lost) to development targets related to health, social and economic effects (e.g. morbidity, livelihood security, economic investment and growth). The core metric is one where the reduction in mortality and morbidity are measured together with a reversal of the loss of livelihoods. The coverage in this transition can be defined, for instance, by the close interdependence between primary production systems, subsistence livelihood strategies, climatic conditions, food security and income generation. In this sense, the impacts of climate extremes, such as droughts, floods and heatwaves, are measured not only by how much is lost but also by the effects on development and livelihoods of people that depend on primary production for their subsistence. It is important to bear in mind that climate factors are not the only factors that stress subsistence systems. Issues of markets, subsidies, access and cultural norms add to the challenge of assuring food security and alleviating poverty. To facilitate interactions and interplay between development and adaptation, both bottom-up and top-down approaches are needed. Linking adaptation and development requires an understanding that off-farm income is critical to livelihoods and overall adaptive capacity.

Development to resilience

The development to resilience transition starts with the recognition that entitlement negotiations and good governance are essential departure points for sustainable development strategies. From embedding adaptation strategies in development, the next transition is to a development pathway that is resilient to a wide range of threats and events, while protecting the poor and most vulnerable. The key characteristics of enquiry are to improve coping mechanisms across a range of traditional and modern adaptation technologies, together with an analysis of community and socially centred bounce-back structures that ensure recovery and continuation of the development trajectory. Validation of the change to a development to resilience paradigm requires evidence that the negative impacts of adverse weather events and climate trends have been significantly reduced.

Development and, in particular, poverty alleviation seeks to reduce the adverse effects of the impacts of variable events by building resilience. Resilience building focuses on improving coping mechanisms and the capacity to recover from disruptive events. This is also termed 'bounce-back' ability. As diversity is the key to building resilience, bounce-back ability is achieved most successfully when both natural biological and social systems' diversities are maintained and enhanced.

Together these processes will help building livelihood capitals and entitlements, but the processes must be realised through negotiation. Negotiation should be seen as transparent and be led by the recipient. Imposed solutions will not work.

Resilience building requires a positive feedback process that reduces impact. Moreover, building bounce-back ability needs appropriate information sets; knowledge of the range, effect and cost of adaptation technologies (both modern and traditional); and access to technologies and recognition that technology, in the broadest sense, changes relations between people and between people and nature.

An enabling and learning environment for knowledge-based activities is fundamental to promote social resilience across a range of scales. Different settings can be more or less conducive to effective learning. Learning requires reflecting upon experience and considering an individual's values and interests in the process of cognition and action. While 'single-loop' learning increases the skills of an individual in an activity, 'double-loop' learning begins to question the framework of assumptions and beliefs. It is this latter learning process that can be an instrument for change, and change can enable a paradigm shift. Reflection and an enabling-learning context can allow for emerging knowledge and new understanding. This builds social resilience. Table 5.1 highlights the change in understanding/structures needed to inform adaptive management for the planning of a new resilience paradigm and Figure 5.2 highlights the temporal dimensions of the shift to resilience.

The three main points sustaining this paradigm shift shown in Table 5.1 are:

1. an understanding that the new paradigm is a dynamic process that has the quality to change and evolve over time;
2. the move from a top-down to a more bottom-up (participatory) approach is needed; and

TABLE 5.1 Changes needed for a new resilience paradigm

From	To
Isolated event	Development process and pathway
Risk is not normal	Risk is expected
Centralised response	Participatory adaptive capacity
Low accountability	Transparency and negotiation
Status quo restored	Transformation

Learning processes

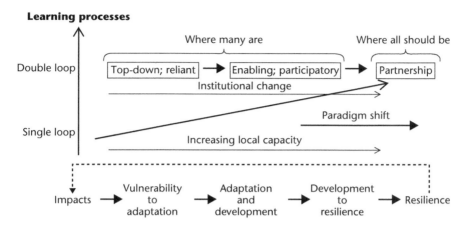

FIGURE 5.2 Temporal dimensions of the adaptation continuum

3. the recognition that one perspective is not more credible than the other, but all perspectives need to be integrated into an enabling framework.

The first point recognises that development can never be risk neutral and that all technological changes have risks associated with them. As such, it acknowledges that risk is 'normal' and is part of the development process as opposed to isolated events.

Most importantly, this point recognises that building resilience to climate change is a process that can change and evolve over time. This means that coping mechanisms should not necessarily seek to restore the status quo of a system but should develop the capacity to adjust to new futures and new kinds of risks. One of the challenges that increased climate variability brings in the short to medium term and climate change in the medium to long term is the challenge of 'produced unknowns'. Although the likely outcomes of a rapidly changing world driven by a shifting climate are broadly known, the actual outcomes cannot be predicted with any precision. This is particularly relevant at the local level where impact science has only produced a broad brush or sector-focused output that is restricted by time. Responding to 'produced unknowns' is a challenge that can only be addressed through strengthening coping capacity in ways that enable it to be flexible and adaptive to the variable challenges it will encounter. For example, the focus on environmental management that Colombia adopted in the NCAP project incorporated this thinking; recognising that coastal natural system resilience is a dynamic process and mechanisms to enhance it and reduce potential impacts will probably change over time.

The second point supports the shift in understanding resilience from an outcome-oriented perspective, which is essentially a top-down and centralised approach that often lacks accountability, to adopting a process-oriented approach that allows for participation, learning and bottom-up processes. The process-oriented approach has its focus not on needs and vulnerabilities but on existing resources and adaptive capacities. The Bolivia NCAP project explored this paradigm shift by adopting a participatory approach to developing adaptation measures based on local perceptions and knowledge, while supporting organisational and social learning to build adaptive capacity first in local communities and then at the national level.

The third point acknowledges that while this paradigm shift is the key in resilience building, it is important to keep in mind that both bottom-up/process-oriented and top-down/outcome-oriented practices are necessary in the process and need to be complementary. Basically, a top-down enabling framework that encourages bottom-up resilience building is the most effective framework.

In short, the underpinning of resilience planning for adaptation includes sustainable development, risk avoidance, least cost intervention, organisational and social learning, and exploring environmental unknowns and tipping points that lead to catastrophic change that moves systems beyond the limits in which resilience can affect a recovery. Resilience planning should be normalised as part of the development process as an issue of social justice. In that sense, it must not be considered an add-on effort but integral, and the impacts of resilience planning must be measurable.

The adaptation continuum framework

The shift from an impact science to vulnerability is a case of adding the social perspective, but the move from vulnerability to adaptation and development in the NCAP studies is one in which social perspectives are understood as dynamic actor–network processes in addition to traditional vulnerability analysis, often based on biogeophysical indicators. It is this shift in perspective that places people at the entry point and prompts the process to integrate socio-economic development and adaptation to biogeophysical impacts. This process will ultimately lead towards building resilience that requires a paradigm shift.

The complexity of the scale of action in the adaptation continuum can be better understood when analysing the sets of information required for the process. To assess impacts, biogeophysical data sets are necessary, whereas vulnerability analysis requires the addition of social data to inform the system. The key characteristics of problem statements of vulnerability, which occur on different scales, are different from impact statements that are defined for specific places and scales. The impact of climate change is conditioned by the variability of vulnerability across space, social groups and economic conditions. Social mapping of vulnerability reflects how vulnerability can be simultaneously constructed at different scales and across time.

It is important to draw adaptation strategies from a wide range of traditional and modern interventions rather than take interventions from a single impact analysis that implies universality to adaptation that is not available. Building adaptive capacity requires moving forward to consider actor–network dynamics. In this regard, the integration of adaptation and development needs to be informed by data on economic and institutional processes. Finally, moving from development towards resilience requires data that provides insight into coping mechanisms and a system to measure the positive feedback process of resilience that reduces climate change impacts. The details of the NCAP studies are important because they generate a wider set of methodological conclusions.

The following framework in Table 5.2 summarises the adaptation continuum features derived from the NCAP study.

General lessons learned and further steps to take

The overall conclusion of the adaptation targets investigation under NCAP was that target setting can help focus adaptation efforts on priority sectors/areas. It can also facilitate monitoring of the effectiveness of managed responses to the adverse impacts of climate change. Country projects in Bangladesh, Bolivia and Mongolia demonstrated that the process of setting targets facilitates the identification of priority sectors, regions and locations and provides ways to monitor the effectiveness of response measures.

Also, vulnerability and resilience assessments conducted by the country teams to identify capacity deficits and to set adaptation targets contributed to defining specific actions needed to build resilience. For example, Mongolia set targets for disaster preparedness in the livestock sector that should demarcate the specific actions needed to reach those targets and move forwards towards resilience. Likewise, Bangladesh proposed targets that form the basis for specific actions to improve sanitation facilities, increase food security and improve health conditions.

A further conclusion is that defining adaptation targets and metrics may also help to prioritise sectors, regions and locations for adaptation investments. This

TABLE 5.2 The adaptation continuum framework features

Process impact	V to A process	A and D perspective	D to R paradigm
Metric	Lives lost	Mortality and morbidity reduced	Livelihoods, entitlements and governance
Key characteristics	Problem definition	Pre-disaster planning integrated sectoral analysis	Improved coping mechanisms bounce-back structures
Models of good practice	Integrated natural resources management	Early warning and climate risk monitoring systems	Cross-sectoral analysis entitlements equity
Information system management	Biophysical and social data social mapping	Data on economic and institutional processes	Data on coping mechanisms, existing capacity, system to measure reduction of impacts

would be a natural progression in the case of the Bolivia and Mongolia projects, which focused on specific vulnerable regions and sectors, respectively. In both cases, once implementation of measures associated with targets commenced, stakeholders would be compelled to determine the financial resources needed for specific problems in the identified areas.

All the teams involved in the NCAP adaptation target investigation agreed that the approach shows promise and suggested that pilot projects would help to gain on-the-ground experience. The adaptation target investigation and the vast wealth of useful studies carried out in 14 countries across three continents during the 5 years of the NCAP all show promise.

Addressing climate adaptation, however, remains a challenge because any policy measures aimed at solving the climate problem will be effective only if they are integrated within a wider set of strategies to promote the efficient and equitable use of resources, good governance, investment and income growth. Meanwhile, developing countries typically lack the institutional capacity and financial means to carry out the range of activities that are needed to formulate, implement and evaluate such policies. Nevertheless, such countries contain many talented and committed individuals. This means that, with international support and cooperation, developing countries can carry out reputable science, as is demonstrated by the high quality of work produced by the NCAP country teams.

Perhaps the most important lesson from NCAP is that any attempt to utilise development planning to create resilience will require not only increased capacity but also greater political will. In short, stronger partnership and cooperation is needed to create an enabling environment that recognises the role of all in resilience building.

Thoughts on adaptation

From an unpromising start, the debate within UNFCCC on adaptation has become increasingly important. The problem of separating countries with genuine adaptation concerns from those concerned about the impact of response measures hampered early discussions. A further problem area has been the lack of agreed definition of adaptation with views ranging from a process that would happen anyway as part of societal development to a problem distant in time through to a matter of urgency that needed immediate action. UNFCCC does not define adaptation, though it is defined by IPCC in its *Third Assessment Report* and its *Special Report on Managing the Risks of Extreme Events and Disasters to Advance Climate Change Adaptation*.

> **Adaptation** – Adjustment in natural or human systems in response to actual or expected climatic stimuli or their effects, which moderates harm or exploits beneficial opportunities. Various types of adaptation can be distinguished, including anticipatory and reactive adaptation, private and public adaptation, and autonomous and planned adaptation.
>
> (IPCC 2001)

The costs of adaptation are likely to be high, running at several billions a year for developing countries alone. Ensuring that climate change is mainstreamed into development policy and international agreements is crucial. Meeting international goals, such as the MDGs, will become more difficult unless adaptation measures are implemented. It is of equal importance that investment projects from whatever source are both 'climate proofed' and 'climate friendly'.

Adaptation is not just adjustment to an average climate condition but is a response to reduce vulnerability to extremes, variability and rates of change at all scales (IPCC 2001). This definition reflects the variety of views of adaptation ranging from an ecological concept in UNFCCC to a series of actions and more recently to a synonym for development (Schipper 2006).

Linking climate change adaptation and development is unavoidable. Over a billion people are surviving on an income of less than a dollar a day. Poverty, however, is more than a low-income indicator. Those living in poverty lack instrumental and substantive freedoms and are often forced to survive by any means possible (Sen 1999). Daily survival in marginal areas poses a threat to human well-being (Kirkby and Moyo 2001). Poverty means that livelihoods are unsustainable in the short term. It is therefore somewhat naive to assume that people living in poverty will, or are capable of, changing their livelihood strategies solely in response to the threat of impending climate change.

In order for people, including those living in poverty, to meaningfully understand and address the impacts of climate change, it is important to realise that climate change and variability are an additional burden on poor people, and usually not the only or most significant. The sustainable livelihoods framework (see Figure 5.3) provides a useful depiction of where climate change can potentially exacerbate the range of stresses on people living in poverty.

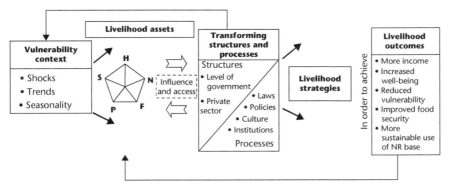

(H = human capital; N = natural capital; F = financial capital; P = physical capital; S = social capital) © Crown copyright 2011

FIGURE 5.3 The DFID sustainable livelihoods framework
Source: DFID (2000).

The framework describes livelihood assets and their relation to wider socio-economic, geopolitical and biophysical processes. Human assets include health, education and skills; financial assets include income, access to finance and insurance; physical assets include shelter and other local infrastructure such as roads or hospitals; natural assets include the means of primary production as well as ecosystems; and social assets include access to groups through family or community. It follows that if people have improved access to livelihood assets they will have more ability to influence structures and processes so that these become more responsive to their needs (Ashley and Carney 1999). Climate change has the potential to further reduce access to the entire range of livelihood assets.

Climate variability is an integral component of people's vulnerability context as depicted in the sustainable livelihoods framework. The IPCC definition of vulnerability is:

> Vulnerability is the degree to which a system is susceptible to, and unable to cope with, adverse effects of climate change, including climate variability and extremes. Vulnerability is a function of the character, magnitude, and rate of climate change and variation to which a system is exposed, its sensitivity, and its adaptive capacity.
>
> (IPCC 2007b: 883)

The use of the term vulnerability is significant because it implies a human ethical dimension not found in the use of the term sensitivity (Downing and Lüdeke 2002). Vulnerability and risk form chronic and cumulative burdens in situations, particularly where traditional coping strategies have eroded or collapsed. The impacts of the maladaptive livelihood responses frequently include heightened mortality (death) and morbidity (ill-health), negative effects on the economy and development, increased stress on environmental resources, and a large diversion of resources from other pressing needs due to environmental degradation (Miller *et al.* 2008; Holling 2001).

It is of crucial importance to understand the vulnerability context of people's livelihoods, rather than suppose we are uniformly 'vulnerable' to climate change as a society or region. If the parameters of analysis include the pre-existing vulnerability context of people who live in poverty, then there is more scope to address their immediate survival needs, as well as any future adaptations that may be necessary in the face of a changing climate.

Adaptation funding

Funding is crucial for adaptation measures. As discussed earlier there is some doubt about where funds for the GCF will come from and the projected figure of USD 100 billion per year by 2020 is unlikely to be sufficient.

In our world nothing, apparently, is really taken seriously until we know the costs. Climate change is no different. Climate change is complex. Reducing the risk of climate change has two different, but related, aspects. The first, mitigation, is about finding ways of reducing GHG production to prevent a future catastrophe or, at least reduce the severity of impacts. This is about shifting or adjusting energy system development from a fossil to a renewable base – energy system adaptation – a fortunate juxtaposition because easily accessible fossil fuels are getting ever scarcer. The second, adaptation, is about protecting us from current adverse events and adjusting to changing realities. This is about building a new world or human adaptation to new climate conditions. Given that level of uncertainty, it is hardly surprising that estimates of how much this will cost are problematic. Additional to this is human suffering. How many will die? By how much will quality of life be degraded? Over what timescales? And for whom? Can we really allow decisions to be taken by governments who appear to want to know the cost of everything but the value of nothing?

Changing the direction of energy system development – the mitigation challenge – is a bit like changing the course of a super-tanker. It is slow. At first nothing appears to happen. Existing energy infrastructure is expected to be in place and producing for many years, depending on when it was commissioned. Infrastructure being constructed today could be around for another 30–50 years or more. Many renewable technologies are still in the development phase and there are serious questions about whether or not sufficient renewable infrastructure could be brought on stream in time for serious emissions cuts to be made.

The numbers are huge. To support a global transition to more efficient, low-carbon energy systems, the IEA estimates that USD 10.5 trillion (with USD 8.3 trillion in end-use) is needed by 2030. With such effort, global CO_2 emissions would decline after 2020 and by 2030 they should be lower than today's level (IEA 2009).

Despite considerable investment in renewable technologies, the sector is dwarfed by fossil fuels, which account for 78 per cent of global energy supplies; wind/solar/biomass/geothermal power generation account for 0.7 per cent; biofuels 0.6 per cent; biomass/solar/geothermal hot water and heating 1.4 per cent; and hydropower 3.2 per cent (REN21 2010). Energy is a crucial aspect of sustainable development but whether we have the political will to adapt the energy system to renewable resources is another question.

Climate adaptation is now receiving serious attention by global policymakers. This is because in earlier negotiations it was thought that adaptation would be an issue for poorer nations – not something that would really affect us (by us we mean the developed world). There was also a fear that the costs for adaptation would mean less funding for mitigation, which was the larger concern of developed countries. Now we realise that we will feel both direct and indirect effects as the planetary climate system re-adjusts. It is likely to be painful and it will certainly be costly. But how costly? And who should pay?

Some work has been done in estimating what the costs of adaptation might be. Studies by Stern and the World Bank on the costs of adaptation, published in 2006, suggested costs between USD 4 and 91 billion per annum (Stern 2006; World Bank 2010b); however, estimates were based on climate-proofing investments – Foreign Direct Investment (FDI), Gross Domestic Investment (GDI) and Overseas

Development Aid (ODA) flows. Studies in 2007 from Oxfam and UNDP estimated costs between USD 50 and 109 billion (Oxfam 2007; UNDP 2007). These estimates were based on the World Bank method but broadened to include NGO projects and poverty-reduction measures. In 2007, UNFCCC estimated that the total funding need for adaptation by 2030 could amount to between USD 49 and 171 billion per annum globally, of which between USD 27 and 66 billion would accrue in developing countries (UNFCCC 2007b). More recent analysis suggests that these figures underestimate the costs of adaptation. One important issue to note is the sector for which estimates are made. Adaptation is complex. Adaptation studies need to include a number of sectors, such as agriculture, water, ecosystems, coastal systems, human health and infrastructure, all of which are necessary aspects of human well-being. The UNFCCC study, for example, did not include mining, manufacturing, energy, retailing and tourism. An analysis of the UNFCCC study by Parry et al. (2009) suggests that in some sectors estimates may need to be two or three times higher, and this is before the interdependencies between and across sectors are fully understood.

A 2010 study by the World Bank into the infrastructure costs alone suggest an annual spend between USD 75 and 100 billion (World Bank 2010a). One of the problems is deciding on what areas and sectors to protect. This implies that the cost of a concerted effort is likely to be much higher that current published estimates. Given that the recent World Bank study only focuses on infrastructure it is not unreasonable to assume that a comprehensive approach is likely to exceed USD 100 billion per annum (World Bank 2010a). This approximates to the total current global annual figure for ODA.

It is unlikely that adaptation will avoid all damages and there will be some residual damage – the amount will depend on willingness to spend. Compounding this problem for modellers is that there is no agreed greenhouse stabilisation level or timescale. Modellers can only best guess what agreement global leaders will make and how rigorously they will adhere to those targets in estimating what the costs will be, and though they will be high, perhaps it is time to reflect more on the fact that billions upon billions of dollars of investment are routinely made as a result of 'uncertain' economic projections, such as GVA projections. Why is it that governments seem to accept economic theory as fact (Urry 2011)?

The costs outlined only account for dealing with physical infrastructure. The 'soft engineering' associated with pre-disaster planning is not even addressed. In the United States, the costs of soft engineering to address future disaster are reckoned to be at least the equivalent of 20 per cent of hard engineering; in essence an additional USD 20 billion.

Across the world, there is general recognition that small-scale disasters cumulatively cost more than large-scale disasters and that small-scale disasters are often beyond insurance markets, leaving especially the poor to bear the brunt of the problem. Munich Re reported USD 200 billion insured losses that were related to climate change (Wilson and Felsted 2008). It is reasonable to assume that real losses amount to double this figure on an annual basis.

The costs of climate-related losses amount to USD 125 billion per annum according to the Global Humanitarian Forum; in addition, some 300,000 lives are lost every year along with countless numbers of lives that have been degraded because of climate change (Global Humanitarian Forum 2010). It is virtually impossible to put a value on such losses and suffering, but for those that need to know the numbers – at least USD 120 billon per annum for adaptation and USD 125 billion of economic losses – a little short of USD 250 billion per annum in total. This does sound a lot, but we should note that the IMF estimates that the costs of the global credit crunch are USD 10 trillion, and in 2006, Europe spent over euro 60 billion on cosmetics and toiletries (Global Insight 2007).

Post-normal adaptation

We have argued that we need a new approach to adaptation and an iterative approach that will allow any new knowledge from climate science to be incorporated into this iterative process. So would a post-normal approach to risk management offer anything different? Figure 5.4 sets out a framework for such a process. For illustrative purposes, Figure 5.4 assumes a relatively small area, such as a city.

The knowledge base represents the area where dialogue takes place within an extended peer-review group. There will already be considerable knowledge about

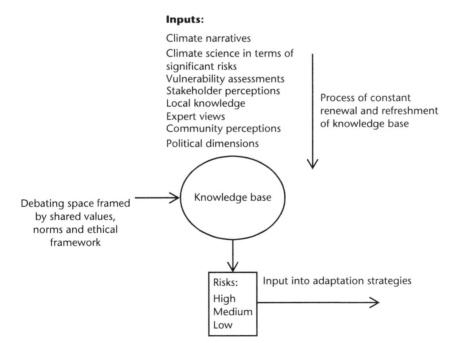

FIGURE 5.4 Post-normal risk management

the area derived from the public and private sectors and local knowledge. Although knowledge of what might happen in terms of climate extremes may be scant, there could be local memories as well as experience to draw on through climate narratives. Governance of this system would be complex and it would be necessary to predicate that on a system of shared values, norms and ethical standards. The method for reaching decisions should not lead to situations of browbeating. When genuine differences arise, further work should be done. Outputs will be a ranking of climate risks that feed into an adaptation process to formulate robust hard and soft measures that maximise co-benefits.

This approach differs from many in use today, which tend to be top-down and expert driven. For example, the United Kingdom has a legal requirement to produce Community Risk Registers where the risk of an area is determined by local government and the emergency response services. There is little doubt that the compilation of the register is done in a very professional manner, but there is no input from the local community or other stakeholders. Furthermore, there is little evidence of any significant impact on the planning system, leaving scope for maladaptation.

Iterative climate management

The IPCC has stated in its Fourth Assessment Report that both mitigation and adaptation should be subjected to an iterative risk management approach (IPCC 2007a). There are examples of countries that have adopted similar approaches, for example the United Kingdom Climate Impact Programme (UKCIP) has adopted a similar approach for its risk management model. It seems clear that adaptation efforts will benefit from iterative risk management strategies because of the complexity, the uncertainties and the long timeframe associated with climate change.

There are a number of approaches and pathways to a sustainable and resilient future, but limits to resilience will be faced when thresholds or tipping points associated with social and/or natural systems are exceeded. This will pose severe challenges for adaptation and has been illustrated by the Copenhagen Diagnosis, which discussed a number of possible areas where thresholds could be breached, tipping points reached and adaptation choices and outcomes determined by the local. Development is uneven with different areas at different stages, determined in part by existing capacity and access to resources. Additionally, an area may experience multiple interacting processes. Choices and outcomes for adaptive actions to climate events must reflect this diversity. This means that there will be trade-offs between competing prioritised values and objectives. It is possible that new development visions will emerge over time as new knowledge is gained. Iterative approaches allow development pathways to integrate risk management so that diverse policy solutions can be considered as risk, and its measurement, perception and understanding evolve over time (IPCC 2007a).

The key factor that will determine the effectiveness of a post-normal approach to risk management, assessment and adaptation is a willingness to embrace changes in thinking. Many methods used for risk assessment and management are based on single-loop learning. This will have to change. We must question and interrogate the methods used in conventional risk management and assessment because there is little data on which to build a model. We must learn from any extreme event that occurs and apply that learning to new post-normal risk management and assessment techniques, however, there are a number of problems that need to be addressed.

Knowledge gaps

We can see from the evolution within the IPCC reports that knowledge and understanding of how human activities have influenced the climate system has grown. But there are still gaps. Continual research is needed to reduce uncertainties associated with climate change and new knowledge needs to be continually fed into the design of adaptation and risk management strategies to improve their effectiveness.

Implications for sustainable development

Disequilibrium will continually drive incremental changes but non-linear events will require transformational changes in order to reduce risk from climate extremes. As argued in this book, the current status quo interpretation of sustainable development, predominantly by the developed world, will need to change. Wildfires and droughts in the United States and storms and flooding in Asia will require change, but if, in the longer term, sea levels surge then this presents a whole new problem because many of our cities are in coastal areas. Do we defend or relocate?

Structural change

Development drives change with the aim of improving efficiency within existing technological, governance and value systems. Transformation, on the other hand, will require either some fundamental changes or a complete rethink. Rethinking approaches will need to embrace new forms of social learning; avoiding change would be disastrous.

Transformation

When non-linear disruptions require a transformative response, there will need to be a greater emphasis on adaptive management and learning. The capacity to do so will depend on the determinants of adaptive capacity. This will vary from country to country but, in general, the greater the amount of the determinants of adaptive capacity a country possesses then the less vulnerable a country will be and the better able to adapt sustainably to climate extremes. Figure 5.5 maps the determinants of adaptive capacity for higher and lower income countries and Table 5.3 summarises the differences. In general, vulnerability is often concentrated

110 Adaptation

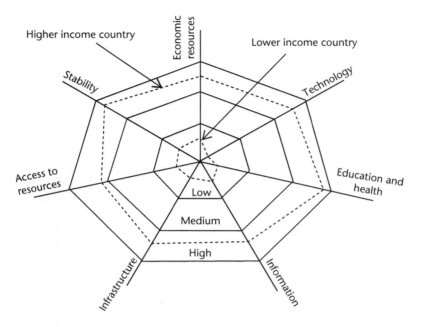

FIGURE 5.5 Heptagon of the determinants of adaptive capacity

in lower-income countries or groups; however, this does not mean that higher-income countries or groups are invulnerable to climate extremes.

Table 5.3 does tend to show extremes but it is possible to discern that countries will have different levels of determinants. Mapping them onto the heptagon of determinants, shown in Figure 5.5, will produce a visual image of where the gaps are. A prerequisite for sustainability in the context of climate change is addressing the underlying causes of vulnerability. This includes structural inequalities that create and sustain poverty and constrain access to resources. This will involve enhancing adaptive capacity before efforts to integrate DRR and adaptation into social, economic and environmental policy domains. Poverty reduction is a precursor to effective climate risk reduction and adaptation. The most effective adaptation and DRR actions are those that offer development benefits in the relatively near term, as well as reductions in vulnerability over the longer term. But poverty reduction will be the first step.

Of course the picture is never quite as clear-cut as suggested in Figure 5.5. Figure 5.6 illustrates what the reality is more likely to be. Many developed countries are mired in status quo thinking with regard to sustainable development and lack the political will to introduce change. As discussed in other parts of this book, it has been the developing countries who have led the charge with respect to adaptation, driven partially by the recognition that they will suffer disproportionately from a problem they are not responsible for and partly because the Climate Convention recognises differentiated responsibilities and, to date, the developed world has not

TABLE 5.3 Determinants of adaptive capacity

Determinant	Higher income countries	Lower income countries
Economic resources	In general, economy based on services and trade across the global. High per-capita incomes. Access to financial resources to fund adaptation.	In general, economy based on agriculture. Some examples of resource rich countries with large export capacity – often this does not benefit local people. Some cases where Foreign Direct Investment in manufacturing either exploits cheap labour and/or resources. Little capacity to fund adaptation.
Technology	Usually technologically advanced with access to manufacturing capability and technological capacity that can help adaptation.	Usually not technologically advanced. Typically reliant on technological imports from technologically advanced countries.
Education and skills	Mainly state funded or supported education with high standards. In general a well-educated and skilled population that enhances adaptive capacity.	Often poor, only offering the basics. Many examples of people from wealthier backgrounds accessing higher education in other countries.
Information	Diverse media (television, radio, print) that are not controlled by the state. Extensive means of personal communications. Expectation that population has a right to be informed. Media has global reach. A sophisticated information communications infrastructure. Usually good access to latest climate science. May not have sufficient or timely information available in terms of adaptation.	Often basic, if any at all. Often state controlled/influenced and used as a means of propaganda. Will rely on other countries/agencies for climate data.
Infrastructure	Usually well established throughout the country with high technology links such as internet connectivity. Some areas may not have adaptation measures in place to deal with climate extremes such as heavy precipitation.	Usually very poor and in some cases hardly any at all. In many cases low levels of connectivity. Lack of infrastructure could hamper implementation of adaptation strategies.

Continued

112 Adaptation

TABLE 5.3 Continued

Determinant	Higher income countries	Lower income countries
Access to resources	Access to resources such as health care, education and social support systems is good. Amount of natural resources can vary – usually intellectual/knowledge/skills base regarded as the key asset. Uses economic and political influence to access resources globally. There are examples of inequitable distributions of wealth with some communities with high levels of deprivation.	Often poor with little access to high quality medicines or health care. Traditional support systems based on the extended family, often little or no welfare system. Aid/welfare programmes often run by NGOs. Often rich in resources, minerals and biodiversity. Often poorly exploited or exploited under the control of developed world organisations/companies.
Stability	Stable multi-party democracy with established institutions. Usually have well-established law enforcement and a legal system that incorporates individuals' rights, rights of expressions and equality.	Often dictatorial, corrupt, closed. Legal framework may be in place but little access to system for poor people. Can often be used to restrict or limit individual freedoms. Police can often be corrupt or used as a means of enforcing the will of the state.

provided the funding. Figure 5.6 also suggests that if the other determinants could be enhanced, then it is more likely that the developing countries will be amenable to transformative adaptation strategies.

In summary, progress towards resilient and sustainable development in the context of changing climate extremes can benefit from questioning assumptions and paradigms and stimulating innovation to encourage new patterns of response. Successfully addressing disaster risk, climate change and other stressors will involve embracing broad participation in strategy development, the capacity to combine multiple perspectives and contrasting ways of organising social relations. We could start by ensuring we meet the MDGs.

Getting it done

The IPCC acknowledges in its Fourth Assessment Report that there is little agreement of the likely changes in seasonal rainfall over large parts of Africa, Asia and South America (IPCC 2007a). Such uncertainties for populations that are vulnerable to variations in precipitation are worrying. Wilby and Dessai (2010) point out that the top-down or scenario-led nature of climate modelling within UNFCCC means that there is uncertainty as to whether or not global impacts can be translated down to the local level. There have been efforts aimed at producing regional

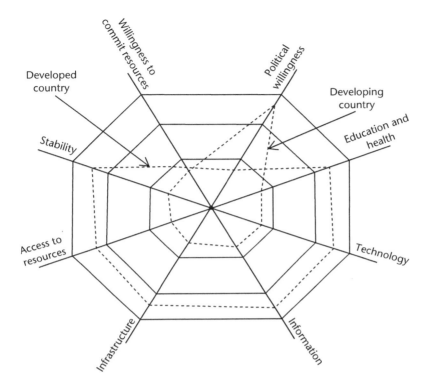

FIGURE 5.6 Octagon of willingness

scenarios that can be translated into local impacts, which can then be used to formulate adaptation strategies, but, as shown in Figure 5.7, each step in the downward cascade produces new uncertainties.

Although this is the method used by IPCC for downscaling, the uncertainties generated are not helpful in formulating adaptation strategies, particularly because an impact focus does not address vulnerability. This means that a 'predict and provide' model for adaptation is flawed by uncertainties and risky: get it right – hero; get it wrong – zero! One of the underlying principles underpinning risk management is the safety of people. Using a bottom-up approach focused on reducing vulnerability to climate risk can be very effective and there is evidence that storm shelters have reduced fatalities in communities in Bangladesh (O'Brien and O'Keefe 2010). There is a history of learning from disruptive events, for example building codes and evacuation procedures in Florida as a response to hurricanes – something the authorities in New Orleans failed to learn from. As discussed under social learning, understanding how individuals, households and communities cope provides valuable insights.

From the earlier discussion, we know from the determinants of adaptive capacity that, in general, wealthier communities should be less vulnerable than poorer

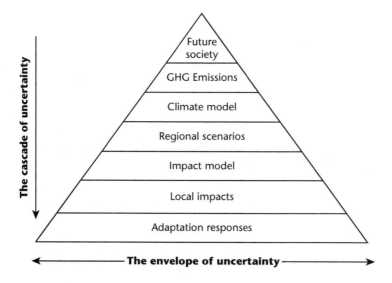

FIGURE 5.7 Adaptation uncertainty

Source: Adapted from Wilby and Dessai (2010).

communities, but such generalisations can mask the range of issues that can exist in a household or community, such as mobility or health problems. Facilities such as schools, hospitals and sheltered accommodation can be very vulnerable. Equally, those driven by socio-economic status and/or by political or discriminatory forces to live in marginal areas, such as floodplains, hillsides or poor quality land, are also very vulnerable. As discussed in this book, vulnerability comprises a range of factors, and vulnerability assessments are an effective way of identifying areas of high vulnerability. Efforts to reduce vulnerability, for example by relocating communities away from a floodplain, are adaptations.

Wilby and Dessai (2010) set out a framework approach for a more robust approach to adaptation. The approach is predicated on access to knowledge of the most significant climate risks and sets out a range of adaptation options (ABCD in Figure 5.8). These are then screened (social acceptance, technical feasibility) to arrive at a set of preferred options. The preferred options can be evaluated against climate models if available or against a narrative of the likely climate future. This robust approach can be characterised as 'low regrets' (see earlier discussion) and, if possible, the choice of measures should be judged against any co-benefits that may arise.

The decision-making challenge

Wilby and Dessai (2010) argue, correctly in our view, that a method for adaptation strategies where climate prediction and risk are not the driving forces but where a shared approach from the bottom-up is the force in the process is more robust;

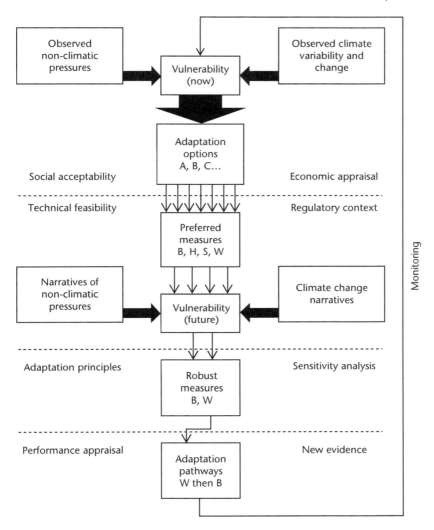

FIGURE 5.8 Conceptual framework for a scenario-neutral approach to adaptation planning

Source: Wilby and Dessai (2010).

however, as presented in Figure 5.8, it is not clear who would be making the social acceptance decision. Would this be expert driven or driven by politicians? The challenge, as we have tried to show in Figure 5.4, is the decision-making space and the governance of that space. Governance is generally regarded as what governments do. Normally, we associate governance with the nation state and the institutions, laws and practices that define how that state functions. Essentially, the sovereign state with its typically pyramidal division of society into families, villages, municipalities, provinces and regions acted as the privileged unit of public governance. Post-Second World War changes in social dynamics have transformed the state. Beck (1992) argues that the Industrial Revolution started a process, and today we live in what is described as the post-modern era where societal changes, such as collectivism, education, female emancipation, rapid technological development and global communications have increased reflexivity (by reflexivity we mean that this is considered to occur when the observations or actions of observers in the social system affect the very situations they are observing). This has resulted in profound change. Increasingly, the role of the state and its institutions, once seen as symbols of authority, has eroded. Globalisation has seen many businesses no longer confined to the nation state. Cross-border trade brings new opportunities and new ideas. Unfettered global markets move vast amounts of money around the globe. Reflexivity has seen the emergence of new social groupings, such as the scientific community where groups of scientists across the world collaborate. It has also seen the rise of global bodies, such as the UN. This, in turn, has led to the development of global conventions, such as the UNFCCC, to address issues that cannot be tackled by a single state. In short, governance has become overlaid with a proliferation of new groupings that is fed by vast flows of information. Economic independence of the nation state is eroding as international bodies regulate global trade and nations become more and more reliant on imported resources, expertise and knowledge. The world is becoming more connected and more complex. Camilleri and Falk (2009) argue that human reflexivity has been at the heart of the process of complexification. They posit that this trend towards what they term 'holoreflexivity' is driven by a desire for self-understanding that encompasses simultaneously the world as a whole and its constituent parts in order to meet our psychological and physiological needs.

What does this imply for governance? Clearly there have been significant changes in governance, and it is clear that a pyramidal or command-and-control model is unlikely to work within the debating space shown in Figure 5.4. Beck (2009) believes that pervasive fear could generate a politics of resistance. This resistance could lead to a 'cosmopolitan vision', a reversion back to a European tradition of cosmopolitan openness to otherness and tolerance of difference. This, he argues, could unite forms of life that have grown out of language, skin colour, nationality or religion with awareness that, in a radically insecure world, all are equal and everyone is different. This could lead to transnational cooperation and pooled sovereignty where actors across the world act to deal with climate risks. This suggests a new form of governance, but Beck is not clear of what incentives or measures would kick-start this process.

Urry (2011), referencing climate mitigation, believes that the challenge will be to use the variety of communications channels to embed the importance of low carbon living within these social worlds. In reality, the energy base of different kinds of society are not yet fully understood and this is an important task for social scientists if we are to move to a low carbon path. From an adaptation perspective, it is clear that we need a greater understanding of vulnerability.

There is a form of climate governance emerging from local initiatives with the state acting through non-state actors, but the boundaries of climate governance are blurred (Bulkeley and Newell 2010). There are questions about the states' ability to harness such initiatives towards a climate target. This is a function of the level of trust the state places in people and local adaptive capacity. The governments of Denmark and Germany have actively encouraged community involvement and ownership of renewable technologies. In contrast, the United Kingdom has not. The difference in implementation is significant.

NGOs play significant roles at the local, regional and national level as lobbyists, educators and community organisers. They can work across national boundaries, for example the International Council for Local Environmental Initiatives (ICLEI) is a network of cities promoting sustainable development including climate mitigation and adaptation; however, the plethora of transnational actors, which can include both public and private actors, raises questions about how, why and where governance is taking place (Bulkeley and Newell 2010).

So what form of governance is likely to emerge? That is almost impossible to predict. The main actions are currently at the grassroots levels: individuals, households, NGOs and local government. One model is panarchy, defined by Sewell and Salter (1995) as an inclusive, universal system of governance in which all may participate meaningfully. Perhaps the real challenge is to develop mutual learning of knowledge and understanding relevant to adaptation where greater transparency on underlying assumptions and values could stimulate a more reflexive, critical and inclusive approach. This could then provide the basis for building a new democratic process for post-normal risk management and climate adaptation.

6
THE CONCEPT OF RESILIENCE

> In order to succeed, people need a sense of self-efficacy, strung together with resilience to meet the inevitable obstacles and inequities of life.
> (Albert Bandura (1925–present))

Introduction

The term resilience has its origins in the Latin word *resilio* and means to jump or bounce back. The resilience perspective as a response to disruptive challenges or contextual change has emerged as a characteristic of complex and dynamic systems in a number of disciplines including ecology, economics, sociology and psychology. The study of resilience evolved from the disciplines of psychology and psychiatry in the 1940s. This focused on efforts to understand the aetiology and development of psychopathology, particularly in studies of children 'at risk' due to, for example, parental mental illness, perinatal problems, inter-parental conflict, poverty or a combination of the above (Manyena 2006). More recently, resilience has been used in studies of ecosystems to describe the capacity of these systems to tolerate disturbance without collapsing into a qualitatively different state that is controlled by a different set of processes (Holling 1973). Within human societies, resilience describes the capacity to anticipate and plan. People are part of the environment and their actions interfere with and modify the environment. This is best described as the production of nature, where people change environmental circumstances to gain benefits, for example the clearance of an area for agricultural purposes.

All life modifies the environment but there is little doubt that humans have made the greatest impact. In the industrialised world, many human activities have detrimental impacts, often to the point where parts of the ecosystem have either been severely damaged or destroyed. There is evidence that the environment can recover from human activities. This is consistent with Lovelock's Gaia theory that the planetary ecosystem acts like a single organism to regulate conditions on Earth to maintain the optimum conditions to sustain life. If a detrimental event occurs, such as a large meteor impact, Gaia is able to use its restorative capacity to adjust to the new conditions and begin the process of recovery. Eventually, Gaia is able to re-establish conditions that are suitable for life (Lovelock 1979).

The other aspect of the human production of nature is that it can expose humans to natural hazards, for example by settling on a floodplain, as well as produce new types of hazards, for example climate change. It is this realisation that human actions can both expose us to and generate new hazards that has driven the interest in resilience for disaster management. Resilience is a function of human–environment interactions and, arguably, by undertaking actions that both enhance the resilience of people and the environments in which they live, it should be possible to reduce the production of hazards as well as being able to respond more effectively to those natural hazards over which we have no control, such as earthquakes. From this perspective, resilience becomes an important concept for sustainable development and development practice.

The concept of resilience

Resilience is the ability to absorb disturbances or events that will bring about change and adjust to those changes but still retain the same structures and methods of functioning. It should be noted that change as the result of a disturbance or event can contribute positively to development. Often these disturbances or events can be rapid and sometimes chaotic, for example the introduction of mobile telephones and the Internet has transformed the ways in which people communicate and share information on a global scale. Essentially, this disturbance to ways in which we communicate and share information has impacted society, but our ability to learn and adjust has meant that society has been able to explore and develop new and innovative ways of using these technologies. Earthquakes, such as the one in Christchurch, New Zealand, are rapid events that can have devastating consequences. In that sense, resilience operates at a range of scales from the local to the regional to the national and the global. It must be recognised that interventions can be either negative or positive.

From this perspective, resilience is a process not an outcome. Resilience is about the ability to respond to change. Change can often be disruptive. Events or disturbances can be characterised as non-linear because they tend to emerge either quickly or creep up as a surprise. The example of mobile telephones and the Internet is an example of a step change. Conversely, climate change can be characterised as a disturbance that has yet to fully manifest itself, in other words a slow onset disturbance. In short, we can describe disturbances as being surprises because they act at different rates. Human society has, by and large, been able to cope with and adjust to interventions. Even events such as the two World Wars in the previous millennia have not halted development. Resilience can be conceptualised as the ability to respond to non-linear disruptions, the ability to reorganise and/or transform and continue to function.

Although the resilience perspective emerged from childhood studies, it was ecology research that led to the use of resilience in disaster management. In ecology, the resilience perspective emerged from predation research in the 1960s and 1970s. Resilience was elaborated as the capacity to persist, and the ecologist Holling proposed that 'resilience determines the persistence of relationships within a system and is a measure of the ability of these systems to absorb changes of state variables, driving variables, and parameters, and still persist' (Holling 1973: 17). The resilience perspective is increasingly used in the analysis of human–environment interactions (Janssen *et al.* 2006). These studies have clustered into a coherent and comprehensive body of research.

The unit of study for human–environment interactions is variously termed the socio-ecological system or SES (Gallopin 1991), a social-ecological system (Berkes and Folke 1998) and a coupled human–environment system (Turner *et al.* 2003). The term human–environment system will be used throughout this book. The scale of human–environment systems can range from the local to the global, constituted by the 'anthrosphere' (the whole of humankind) and the ecosphere (Gallopin 2006: 294). As a tool, this provides huge flexibility. What has emerged throughout the evolution of research into human–environment systems is the linkage between resilience, vulnerability and adaptation and these concepts are applicable both to the biophysical and social realms.

Resilience is seen as adjustments to return to a steady state or basin of attraction, where, without further perturbations, the system would remain. In reality, the world is a dynamic and often chaotic place and it can be envisaged that systems, including human–environment systems, are constantly adjusting. The ability to do so is a function of both the adaptive capacity of the system and its vulnerability to particular types of perturbations.

Walker *et al.* (2004) further defined resilience as the 'capacity of a system to absorb disturbance and reorganise while undergoing change so as to still retain essentially the same function, structure, identity and feedbacks – in other words stay in the same basin of attraction.' Gallopin (2006) states that the concept of basins of attraction, also termed domains, is central to the notion of resilience. They are described as:

> The portion of the state space of a dynamic system that contains one 'attractor' toward which the state of the systems tends to go, and is therefore one region of the state space where the system would tend to remain in the absence of strong perturbations.
>
> (Gallopin 2006: 297)

Disturbances or perturbations can either be internal or external, single or multiple and of short or long duration. In the absence of disturbances or perturbations, the system is likely to remain in the same state or a steady state. Box 6.1 discusses the concept of basins of attraction.

BOX 6.1

BASINS OF ATTRACTION

Latitude: the maximum amount the system can be changed before losing its ability to recover, basically the width of the basin of attraction. Wide basins mean a greater number of system states can be experienced without crossing a threshold (*L*, Figure 6.1).

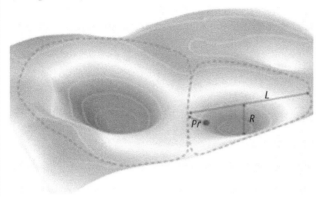

Figure 6.1 Three-dimensional stability landscape with two basins of attraction showing, in one basin, the current position of the system and three aspects of resilience: $L =$ latitude, $R =$ resistance, $Pr =$ precariousness

Resistance: the ease or difficulty of changing the system. Related to the topology of the basin, deep basins of attraction (R, or more accurately, higher ratios of $R:L$) indicate that greater forces or perturbations are required to change the current state of the system away from the attractor.

Precariousness: the current trajectory of the system and how close it is currently to a limit or 'threshold', which, if breached, makes recovery difficult or impossible (Pr).

Discussion

A 'basin of attraction' is a region in state space in which the system tends to remain. For systems that tend toward equilibrium, the equilibrium state is defined as an 'attractor' and the basin of attraction constitutes all initial conditions that will tend toward that equilibrium state. In the real world, conditions constantly change. Using rangeland as an example, it is possible to have more than one basin of attraction.

The various basins that a system may occupy, and the boundaries that separate them, are known as a 'stability landscape'. Figure 6.1 shows the first three components of resilience for a basin in a stability landscape of two state variables.

External drivers, such as rainfall, and internal processes, such as management practices, can lead to changes in the stability landscape (number or positions of basins of attraction), changes in position of thresholds or edges between basins (latitude L in Figure 6.1), and changes in 'depths' of basins, which is a measure of the difficulty of moving a system around the basin – steep sides imply greater efforts are needed to change the state of the system, i.e. the position of the system (resistance R in Figure 6.1).

122 The concept of resilience

> Moving the system around changes its position within a basin relative to the edge (precariousness – Pr in Figure 6.1), or moves it into a new basin (Figure 6.2) where, without the state of the system itself changing, the system finds itself in a new basin of attraction, owing to changes in the stability landscape.
>
>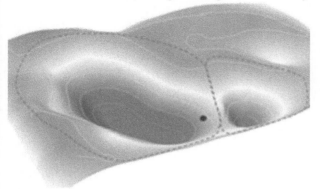
>
> **Figure 6.2** Changes in the stability landscape have resulted in a contraction of the basin the system was in and an expansion of the alternate basin. Without itself changing, the system has changed basins
>
> Source: Walker *et al.* (2004)

From this it can be seen that resilience is the adaptive capacity of a system to respond to perturbations or disturbances. Research by Holling has led to further insights into adaptive cycles. This research sees adaptive cycles as the flow events between four systems, represented in Figure 6.3.

In Figure 6.3, r is termed the exploitation stage. This system is one of rapid expansion where, for example, a species is able to exploit an opportunity. This could represent a case where pioneer tree species are able to exploit an opportunity. The conservation stage, K, is one in which slow accumulation and storage of energy and material is emphasised as the forest builds complexity. As complexity increases, new niche opportunities arise that new species are able to exploit. The release stage, Ω (omega), happens rapidly, for example fire or disease that results in partial destruction of the forest. The reorganisation stage, α (alpha), can also occur rapidly. New species are able to exploit the opportunity and occupy the space made available by the disruption. This then leads to a further cycle of exploitation and conservation where the forest regenerates in a similar pattern to the original forest; however, it does not exactly replicate the original forest but allows new species to exploit opportunities. These changes can also ripple through the surrounding forest causing slight changes and perhaps new opportunities. Providing disruptions do not exceed the capacity of the system to respond, then regenerative cycles will continue. Disruptions such as forest clearance will completely change the state of the system to a point where recovery is not possible. Though this may lead to opportunities

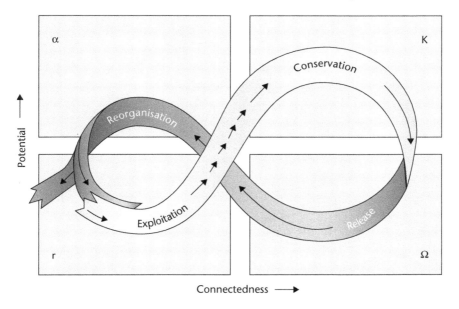

FIGURE 6.3 Representation of four systems and flows between them

Source: Holling (2004).

for those that cleared the forest, it can also lead to unintended consequences, such as increased flood events because forests tend to hold water and release it slowly into watercourses.

This cycle of accumulation and renewal in response to disruptions exhibits resilience. This is shown in Figure 6.4.

This sees resilience as process. During the conservation stage, the increase in complexity acts to increase vulnerability (and consequently lowers resilience) and providing the system does not become so complex that it becomes resistant, i.e. unable to recover from disruptive events, this process of destruction, the rapid back-loop of release and reorganisation, and the slow forward-loop of exploitation and conservation can generate new novelties that are exploited to allow for continual renewal. Important aspects of this concept are scale, which can range from the small to the large, and time where forward-loop processes can take many years whilst back-loop processes can occur quickly. It is the interaction of the two cycles when the rapid, and somewhat chaotic, back-loop generates new opportunities for the slow-moving front-loop.

As stated earlier, this concept can be applied to human–environment interactions. The Resilience Alliance (2011) identifies three characteristics of human–environment systems:

1. the amount of change the system can undergo and still retain the same controls on function and structure;

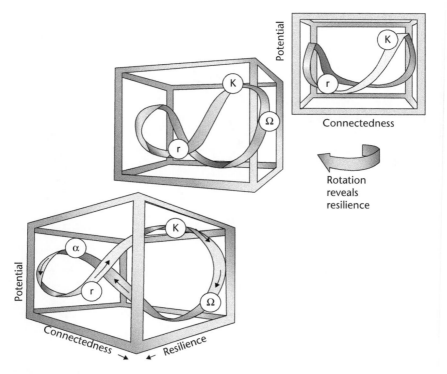

FIGURE 6.4 Resilience: another dimension of the adaptive cycle

Source: Holling (2004).

2. the degree to which the system is capable of self-organisation; and
3. the ability to build and increase the capacity for learning and adaptation.

This can be seen in human developments. Many of our cities have developed from small settlements, often next to rivers. Rivers provide a ready source of water, and land near to a river is usually suitable for agriculture and pastoralism. Rivers are also excellent communication routes allowing the exchange of goods, services and ideas. In the early days, we call this the exploitation phase. As the settlement expands, more skills and knowledge are gained. New forms of governance emerge and the built aspects of the settlement start to become more established. In short, a town begins to emerge. As the town continues to develop, we could call this the conservation phase. Rivers can be hazardous. Low flow could result in drought and high flow in flood. Assume that the town is hit by a flood. This would trigger the release phase where the town's people would respond to the flood itself. The acute event phase and the post-event, or recovery period, would see all the capacity and resources of the community used in an attempt to cope with and recover from the flood. The reorganisation phase would see the accumulated resources, skills and knowledge, along with any lessons learned from the flood episode, used to rebuild and repair any damage. It is in this period when decisions about flood defences or any relocations are likely to be made. Any improvements that mitigate future flooding can be seen as building the resilience of the community. The longer-term recovery would see the town embark on another, slightly different exploitation and conservation trajectory.

Thinking about these concepts enables us to see development processes through this lens. It is clear that many historical cycles can be mapped using the resilience concept, such as the rise and fall of empires. As empires develop, they tend to become more rigid and hence more vulnerable. To an observer, an empire may seem extremely powerful but, as history teaches, rigid hierarchical structures can collapse very quickly. Recent events in North Africa have seen regimes collapse in Tunisia, Egypt and Libya, where seemingly powerful leaders have been quickly removed because they could not respond to demands for change. In short, they were resistant as opposed to resilient. Leaders who have been able to either anticipate demands or react quickly have been able to relaunch along another path. Given the huge range of events or disturbances that impact societies, it is not surprising that the political and economic map of the world undergoes such huge changes.

Similarly, at a small scale, events and disturbances can bring change at the local and regional levels. This includes, for example, industrial and business restructuring in response to market demands and emerging technologies. Post-Second World War, the United Kingdom entered a new phase of economic development where the smokestack or heavy industries were phased out and new enterprises emerged, such as the knowledge economy, specialist manufacture and the service sector. The credit crunch is a reminder that too great a dependence on one sector can have damaging consequences if it is left to its own devices. This illustrates the importance of diversity when building resilience.

It can be seen that these cycles can operate over varied temporal and spatial scales. It is in the chaos of the back-loop where innovation and re-organisation without careful thought can launch a development trajectory that is either unpredictable or has the potential for negative consequences.

Cycles at different temporal and spatial scales can interact. Holling terms this 'panarchy' where 'pan' refers to the Greek God Pan, who represents the vagaries of nature or the world in which we live; the 'archy' refers to governance or rules. The term panarchy was first put forward by de Puydt, a Belgian botanist, who applied the concept to the individual's right to choose any form of government without being forced to move from their current locale (de Puydt 1860). The governance perspective of panarchy has been broadened by others, for example Sewell and Salter (1995) see panarchy as an inclusive and open form of government in which everyone can participate. Sewell and Salter portray Pan as a playful steward of biospheric well-being. Panarchy also emerged in systems theory, an interdisciplinary field of science that studies the nature and processes of complex systems in physical and social sciences, as well as information technology. From an information technology perspective, panarchy is viewed as a transdisciplinary investigation of the political and cultural philosophy of 'network culture'.

126 The concept of resilience

The investigation continues into the fields of world politics (international relations) and political philosophy/theory. From a physical and social sciences perspective, Gunderson and Holling view panarchy as a framework of nature's rules that is the antithesis of hierarchical structures. Gunderson and Holling (2001: 5) describe panarchy as 'The cross-scale, interdisciplinary and dynamic nature of the theory has led us to coin the term panarchy for it. Its essential focus is to rationalise the interplay between change and persistence, between the predictable and unpredictable.' This interaction between systems is illustrated in Figure 6.5.

Figure 6.5 shows how small, faster cycles can affect slower-moving cycles (revolt). Large, slower cycles can control the renewal of smaller cycles (remember). This can be envisaged in the following way. Assume that the middle cycle is an industrial estate. The smaller cycle represents an innovation. The revolt aspect sees the establishment of a new operation to develop and exploit the innovation. The period of release and reorganisation sees activities to support the new development, for example the provision of premises. Assume that the new enterprise is successful and expands output. This can be supported by the slow-moving cycle that can provide resources (physical and human). This interaction sees the development of a new trajectory.

Resilience is the capacity to respond to disturbances. As can be envisaged, there are myriad cycles operating across the planet from the very small to the very large. The interaction between these cycles determines the development path of

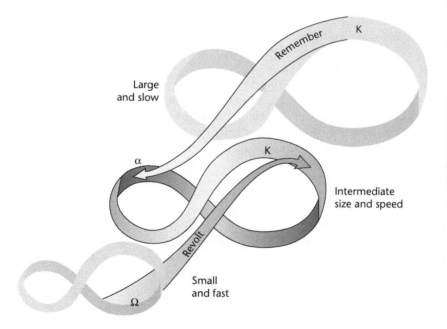

FIGURE 6.5 Connections between three levels of panarchy
Source: Holling (2004).

the planet. Prior to the emergence of people, the planetary ecosystem established a huge variety of life. The fossil record shows that many species have become extinct as different disturbances altered environmental conditions. Despite external disturbances, such as variations in the orbit and spin of the planet and impacts by meteorites, the ecosystem has been able to recover and re-establish conditions necessary for life to flourish. As Lovelock argues, the ecosystem, despite the disequilibrium of the system, acts to maintain conditions for life, but, as Lovelock points out, there are limits (Lovelock 1979). There are concerns that human interference with planetary processes could exceed the capacity for recovery. Accelerated climate change is viewed by many as the single greatest threat to the international aims to achieve a sustainable development trajectory and address the MDGs. Energy is fundamental to life; however, the development of the fossil fuel economy has led to massive growth in energy use. The release of GHGs from energy use (and other human activities) is altering the composition of the atmosphere and warming the planet.

In response to this challenge, the international community has come together, through UNFCCC, to develop solutions. UNFCCC has recognised the need to have sound scientific advice as the basis for formulating policy. IPCC fulfils this role by drawing from peer-reviewed science in a number of disciplines that try to establish the likely impacts of climate change if no further actions are taken, and tries to model the impacts of different policy frameworks. IPCC does recognise that scientific understanding of planetary processes is limited. There is a clear need to deepen and broaden understanding of these processes. Despite these limits, IPCC has presented evidence that human-induced climate change is happening. Though IPCC has been hampered by the ClimateGate scandal, there does appear to be consensus that action is needed to limit the impact of human activities on the atmosphere.

UNFCC has two strategies for dealing with climate change. The first is mitigation, which aims to reduce GHG emissions and hence reduce future risks. The second is adaptation, which aims to reduce the impact of historic and ongoing emissions and hence reduce current and ongoing risks. As discussed in Chapter 1, the emphasis in UNFCCC has been on mitigation. Copenhagen and Cancún have both failed to deliver a successor to the Kyoto Protocol. It seems unlikely that any international agreement of GHG reduction strategies will be made in the near future. This places adaptation at the forefront. Looking at what has happened to UNFCCC from a resilience perspective, the failure to formulate an agreement to reduce GHG emissions means that the conservation phase of the organisation is resistant to change, despite the clear disturbance of IPCC in providing evidence of human-induced change. This resistance is likely to heighten vulnerabilities in terms of risk from climate disturbances and of the global energy system itself because fossil fuel resources are becoming increasingly scarce. There is an urgent need to develop and use new sources of energy.

What is clear is that resilience is an ongoing process that is a function of adaptive capacity and vulnerability, where resilience acts as a counter or antidote to vulnerability. This at the core of the concept of resilience and its focus is not on 'what is missing in a crisis (needs and vulnerabilities) but on what is already in place (resources and adaptive capacities)' (O'Brien et al. 2006: 71).

Resilience, as a concept, has become part of the disaster management, sustainable development and climate change discourses. In disaster management, resilience is variously described as minimising losses and damage (Burby et al. 2000; Mileti 1999), the ability to recoil from adversity (Johnston and Paton 2001), self-righting and learned resourcefulness (Paton et al. 2000) and coping capacity, ability to react or effectively recover (McEntire 2001). The Hyogo Declaration, an outcome of the World Conference on Disaster Reduction (WCDR), states the need to build the resilience of nations and communities to disasters (Hyogo Declaration 2005). The Hyogo Framework for Action 2005–2015 sees building resilience as a key component of planning and programming for sustainable development (Hyogo Framework for Action 2005). The definition of resilience used within the framework is that of the UNISDR:

> The ability of a system, community or society exposed to hazards to resist, absorb, accommodate to and recover from the effects of a hazard in a timely and efficient manner, including through the preservation and restoration of its essential basic structures and functions.
>
> (UNISDR 2009: 24)

The World Summit on Sustainable Development (WSSD) sees DRR and enhancing resilience as core elements of sustainable development. Resilience is also seen in the climate change domain. The UNFCCC uses resilience when referring to the impacts that adverse climate change can have on natural and managed ecosystems and on the operation of socio-economic systems (UNFCCC 1992), but it is more predominantly seen in adaptation where it is described as adjustments to reduce damage or taking advantage of opportunities in response to climate change (McCarthy et al. 2001).

We can see from Table 6.1 that there are a considerable number of resilience concepts. Resilience refers to characteristics of society, social groups and individuals which, following trauma or initial crisis and impact, allow certain sectors and populations to recover with greater facility than others. In general, most concept definitions emphasise a capacity for successful adaptation in the face of disruptions, stress or adversity. It is possible to derive two important properties of resilience from the literature:

1. resilience is an ability or process not an outcome; and
2. resilience is more focused on adaptability than stability.

The focus on ability and adaptability clearly exhibits a strong learning component, pointing to resilience being a learning process that is able to adjust continually to newly emerging patterns.

TABLE 6.1 Resilience concepts

First author, year	Level of analysis	Definition
Gordon, 1978	Physical	The ability to store strain energy and deflect elastically under a load without breaking or being deformed.
Bodin, 2004	Physical	The speed with which a system returns to equilibrium after displacement, irrespective of how many oscillations are required.
Holling, 1973	Ecological system	The persistence of relationships within a system; a measure of the ability of systems to absorb changes of state variables, driving variables, and parameters, and still persist.
Waller, 2001	Ecological system	Positive adaptation in response to adversity; it is not the absence of vulnerability, not an inherent characteristic, and not static.
Klein, 2003	Ecological system	Ability of a system that has undergone stress to recover and return to its original state; more precisely (1) the amount of disturbance a system can absorb and still remain within the same state or domain of attraction and (2) the degree to which the system is capable of self-organisation (see also Carpenter *et al.* 2001).
Longstaff, 2005	Ecological system	The ability by an individual, group, or organisation to continue its existence (or remain more or less stable) in the face of some sort of surprise… Resilience is found in systems that are highly adaptable (not locked into specific strategies) and have diverse resources.
Resilience Alliance, 2006	Ecological system	The capacity of a system to absorb disturbance and reorganise while undergoing change so as to still retain essentially the same function, structure and feedbacks – and therefore the same identity. (Retrieved 10/16/2006 from http://www.resalliance.org/564.php)
Adger, 2000	Social	Ability of communities to withstand external shocks to their social infrastructure.
Bruneau, 2003	Social	Ability of social units to mitigate hazards, contain the effects of disasters when they occur, and carry out recovery activities in ways that minimise social disruption and mitigate the effects of future earthquakes.
Brown, 1996	Community	Ability to recover from or adjust easily to misfortune or sustained life stress.
Sonn, 1998	Community	Process through which mediating structures (schools, peer groups, family) and activity settings moderate the impact of oppressive systems.
Paton, 2000	Community	Capability to bounce back and to use physical and economic resources effectively to aid recovery following exposure to hazards.

Continued

TABLE 6.1 Continued

First author, year	Level of analysis	Definition
Ganor, 2003	Community	Ability of individuals and communities to deal with a state of continuous, long term stress; the ability to find unknown inner strengths and resources in order to cope effectively; the measure of adaptation and flexibility.
Ahmed, 2004	Community	The development of material, physical, socio-political, socio-cultural, and psychological resources that promote safety of residents and buffer adversity.
Kimhi, 2004	Community	Individuals' sense of the ability of their own community to deal successfully with the ongoing political violence.
Coles, 2004	Community	A community's capacities, skills, and knowledge that allow it to participate fully in recovery from disasters.
Pfefferbaum, 2005	Community	The ability of community members to take meaningful, deliberate, collective action to remedy the impact of a problem, including the ability to interpret the environment, intervene and move on.
Norris, 2008	Community	A process linking a set of adaptive capacities to a positive trajectory of functioning and adaptation after a disturbance.
Renn, 2002	Community	Countermeasure to uncertainties by avoiding irreversibilities and vulnerabilities.
Cutter, 2008	Community	Ability of a social system to respond to and recover from disasters.
Walker, 2004	System	Capacity of a system to absorb disturbance and reorganise while undergoing change so as to still retain essentially the same function, structure, identity, and feedbacks.
Rose, 2007	System	Ability of an entity or system to maintain function (e.g. continue producing) when shocked.
O'Brien, 2010	System	Ability to withstand and adjust to disruptions whilst still retaining function.
Masten, 1990	Individual	The process of, capacity for or outcome of successful adaptation despite challenging or threatening circumstances.
Egeland, 1993	Individual	The capacity for successful adaptation, positive functioning, or competence… despite high-risk status, chronic stress, or following prolonged or severe trauma.
Butler, 2007	Individual	Good adaptation under extenuating circumstances; a recovery trajectory that returns to baseline functioning following a challenge.

TABLE 6.1 Continued

First author, year	Level of analysis	Definition
Rose, 2009	Economic	Process by which a community develops and efficiently implements its capacity to absorb an initial shock through mitigation and to respond and adapt afterward so as to maintain function and hasten recovery, as well as to be in a better position to reduce losses from future disasters.
Godschalk, 2003	City	Sustainable network of physical systems and human communities, capable of managing extreme events; during disaster, both must be able to survive and function under extreme stress.

Source: Adapted from Norris *et al.* (2008).

Resilience, vulnerability and adaptability are interlinked, but it is not always clear where the line between the different terms is. Vulnerability and resilience are somewhat generic concepts, and the underlying or driving factors often overlap to make distinctions between the two unclear.

Does vulnerability influence adaptive capacity or does adaptive capacity determine vulnerability? Does decreasing sensitivity enhance adaptive capacity? Does reduced vulnerability always lead to increased resilience?

An analysis by Cutter *et al.* (2008) of relationships between vulnerability, resilience and adaptive capacity grouped the different notions of the terms in global environmental change and (environmental) hazards research. The big difference is in moving from single stressors (hazards) to multiple stressors (global change). Resilience is either perceived as an integral part of adaptive capacity (A), as a main component of vulnerability (B) or as nested concepts within an overall vulnerability structure (C) as shown in Figure 6.6.

In summary, resilience is used as a concept in a number of communities and there is a convergence in the way that the term is expressed: adapt, adjust, change, capacity to absorb or change, recover from and self-organise in response to external disturbances are all common to each school.

Resilience and disaster management

Resilience, conceptually, is the ability to cope with disturbances. It is not surprising that resilience has entered the disaster management field. Disasters can be thought of from a number of perspectives, such as high energy or rapid onset events or low energy or slow onset events; however, all disaster types have a common feature: they bring about change. Although change is a part of the everyday, disasters bring changes that were not expected or are beyond expectations. In this sense, disasters can be thought of in broadly two ways: routine and surprises. Routine disasters can be thought of as those that arise from the everyday such as road-traffic accidents.

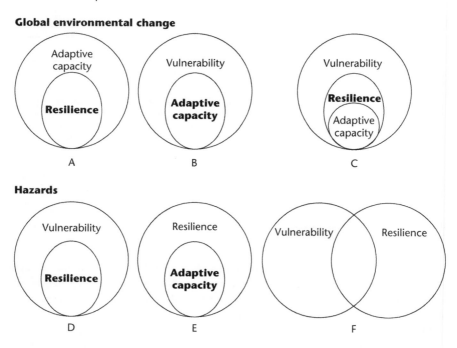

FIGURE 6.6 Conceptual linkages between vulnerability, resilience and adaptive capacity
Source: Cutter et al. (2008).

Between 1970 and 2010, 3.3 million people were killed in disasters, but every year 1.2 million are killed in road-traffic accidents (World Bank 2010c). Fatalities through such incidents as road-traffic accidents can be thought of as part of the everyday; disasters are perceived differently. In the events of a disaster, there is the perception of a lack of control. In fact, many disasters are caused by human action, for example by developing in hazardous areas or developing risky technologies. Table 6.2 shows how disaster events can be characterised.

In the left hand column of Table 6.2 are two disaster types listed as routine. In many respects routine might not be considered an appropriate term. In this context it refers to events that we prepare for, for example drivers are usually trained and require insurance. Roads and highways have rules for drivers and there is a large array of signage and traffic control measures designed to minimise risk. Nonetheless traffic accidents do occur. In the case of nuclear reactors, stringent safety standards are applied and operators are required to have monitoring systems in place as well as conduct drills and exercises to enhance preparedness. Despite such precautions accidents can happen.

In the right hand column of Table 6.2 disaster surprises are characterised. Categories of surprises are shown and it can be seen that they relate strongly to technological development that has unforeseen consequences. These can be described

TABLE 6.2 Characterising disasters

Routine	Surprises
Common (high probability, low consequence) e.g. Road-traffic accidents	Never envisaged
	Envisaged but regarded as highly unlikely
Rare (low probability, high consequence) e.g. Chernobyl	Thought likely to occur but had not happened previously
	Thought unlikely to occur but mistakenly believed to have minor consequences
	Minimata, Exxon Valdez, challenger, DDT, ozone depletion, climate change

as produced hazards that lead to produced disasters. Climate change is an example of a produced hazard that is generating new kinds of risks. Given the lack of an agreed GHG reduction strategy, GHG concentration levels and timetable mean that in the future the nature (scale and frequency) of those hazards will be more difficult to predict, making any kind of risk reduction strategy problematic. Climate change is driving into the unknown.

Use of resilience in disaster management

The use of the resilience concept in disaster management has evolved but it has only relatively recently entered the disaster management lexicon. During the Cold War era, disaster management was practised from a civil defence perspective, with all of the militaristic and command-and-control overtones that would be expected in a defensive posture. With the emergence of the all-hazards approach in the post-Cold War era and a greater emphasis on civil protection, many of the structures of the civil defence approach were incorporated into civil protection structures. Top-down command-and-control structures dominate conventional disaster management in the developed world, but it should be noted that many countries have adopted similar platforms. Major emergencies are usually considered isolated one-time events as distinct from routine emergencies, such as traffic accidents, and the risks are perceived as not being normal. Risk management systems, for example flood warning systems or earthquake response plans, usually involve technology to detect unusual events or sophisticated heat detection systems for finding survivors in the event of a building collapse. For hazardous facilities, there are legal frameworks underpinning management and operating procedures, such as the Seveso Directive. The emergency management system is usually centralised through a government agency, for example FEMA in the United States, or directly in government, for example CCS in the UK Cabinet Office. After an emergency response, lessons learned from the event that can be incorporated into future responses become part of the post-planning activities. The objective of the emergency management cycle is to restore the status quo, for example after the floods in the United Kingdom in

2000, damaged homes were refurbished despite the fact that these areas are likely to flood again.

As discussed in Chapter 4, the emphasis on top-down structures and institutional resilience and the use of militaristic approaches is counter-productive to building societal resilience. A number of scholars began to recognise the need for a more holistic approach to disaster management. Mileti (1999) argued that the technocratic model of disaster management exhibited weakness because of its reactive approach. McEntire and Fuller (2002), in a study of El Niño-triggered disasters in Peru, concluded that in quickly changing environments a centralised command-and-control approach quickly proved inadequate. They argued that a greater emphasis was needed on preparedness.

Lagadec, in an article discussing new approaches to crisis management in the twenty-first century, argues that civil society should be put back in the loop (Lagadec 2005). He suggests that the militaristic approach that sees control as a means of preventing panic and looting is not the model we should adopt, particularly when dealing with unique or unusual events. Lagadec uses the example of the 1988 Quebec ice storms to show that the public behaves in rational and responsible ways, provided they are given the correct information and are seen as part of the process. Research by Quarantelli shows that although some officials may fear that the public might panic, this is often not the case and people do behave rationally in stressful situations (Quarantelli 2005).

There have been a number of efforts to establish a disaster management paradigm that has a holistic approach, one that provides a theoretical framework that guides both researchers and practitioners. McEntire *et al.* reviewed a number of approaches, namely comprehensive disaster management, the disaster-resistant community, the disaster-resilient community, sustainable development and sustainable hazards mitigation and invulnerable development (McEntire *et al.* 2003).

Comprehensive disaster management incorporates mitigation, response and preparedness along with the range of actors involved in disaster management. A major problem with this approach is too great a focus on hazards and that it fails to fully address social, political, economic and cultural factors that contribute to disaster, meaning that this approach is somewhat narrow, despite its claim to be comprehensive.

The disaster-resistant community focuses on assisting communities to minimise vulnerabilities to natural hazards by use of both structural and non-structural methods. The idea is to make communities safe; however, its main focus is on natural hazards, in particular extreme events. This ignores other drivers for disaster, such as social, political and technological factors. This narrowness casts doubt on its effectiveness to act as a holistic approach to disaster management.

The disaster-resilient community approach is criticised because there does not appear to be an agreed definition of resilience and because it has too great a focus on natural hazards. As will be discussed later in the chapter, there is now a sufficiently agreed and comprehensive definition of resilience that recognises the range of hazards society can be exposed to.

The sustainable development and sustainable hazards mitigation approach does have much to offer because of the clear link between development and disasters. This approach is criticised because it appears to focus too heavily on a narrow range of factors, which tends to negate the claim of this approach to being holistic.

Invulnerable development is elaborated as development that addresses vulnerabilities; however, there are two areas of concern. The first is that because of the focus on vulnerability, or rather invulnerability, the concept may be interpreted as there being no disasters. The second is the lack of consensus on the definition of development.

In trying to reconcile these problems, the comprehensive vulnerability management approach attempts to address the shortfalls identified in the approaches outlined above; however, it fails to address how to prepare for produced unknowns. In terms of dealing with what might be termed routine events there is much merit in this approach, but, as discussed earlier, accelerated climate change and increased variability are producing a range of future threats that could have very severe – and unforeseeable – consequences. In addition, there is a lack of political consensus on mitigation within the international community. The lack of a target and timetable means that the future is very open. Even if there were an agreed target and timetable, there would only be a very vague idea of what climatic conditions would ultimately prevail. This uncertainty means that an approach to disaster management focusing on vulnerabilities and livelihoods is needed because we have to ensure that climate risks become part of the everyday and that purposeful adaptation is implemented.

The Hyogo Framework for Action identifies pre-disaster planning and a culture of prevention and resilience as key components for risk reduction at all levels, along with knowledge of societal vulnerabilities as the starting point for actions. Actions are framed around governance, risk identification and reduction, and preparedness (UNISDR 2005). This global declaration, relevant to all nations, recognises the equal importance of both bottom-up and top-down approaches. One of the issues with a top-down focus is that it can become too narrow and therefore give too little attention to the wider range of threats we face, such as climate change (O'Brien and Read 2005; O'Brien 2006). The threats faced are challenging and certain to become more so over time. Preparedness is a concomitant process requiring engagement of all actors. This means that the government must begin to involve the wider public in meaningful and open discussion. Pursuing institutional resilience and focusing on too narrow a range of threats only serves to increase the distance between the government and the governed.

Figure 6.7 illustrates a governance framework for resilience building. This makes it clear that a top-down structure that focuses on institutional resilience is inappropriate if we are to seek a broader societal preparedness agenda. There is a real need to bring together all stakeholders in resilience building if we are to develop effective responses to climate risks.

Broadly, risk management can be thought of as a chain – the closer the links, the stronger the chain. In order to build resilience, all actors need to be close –

FIGURE 6.7 Enabling framework for resilience building

from government to the people. A lack of trust can act to distance a particular actor in this building process, which will ultimately introduce weaknesses. Enhancing resilience involves engagement at all levels. As McEntire states:

> Recognizing the human role in disasters, taking responsibility for action, having a disaster plan, building capabilities to implement the plan, purchasing insurance, and sharing information about recovery priorities are processes that can enhance resilience for an individual, group, community or nation to deal with unique destabilizing events. In this instance, resilience is thus a goal that we should strive to achieve or a quality that we should try to obtain.
>
> (McEntire *et al.* 2003)

This requires trust at all levels; effectively resilience building is a partnership between government and governed, but profound changes took place during the twentieth century that shaped public attitudes. For example, in the United Kingdom, great social and technological changes in the latter half of the twentieth century led to a rise in individualism and the information revolution led to the emergence of a highly educated and increasingly mobile information society, typified by the declaration by Margaret Thatcher that there was 'no such thing as society' (Thatcher 1987).

The corollary to this was the transfer of loyalties from institutions and structures to individual values or individualism. Beck terms this reflexive modernity and argues that the Industrial Revolution, the first modernity, saw many radical changes in everyday life, yet it was still based on traditional social structures, particularly family and gender (Beck 1992, 1995). The latter half of the twentieth century, the second modernity, saw further changes, for example women entering the workplace, the shift from full-time to part-time employment, the erosion of lifetime job security in both blue-collar and white-collar occupations and changing family and social structures, which began to modernise the foundations of the first modernity, making it reflexive. This latter part of the twentieth century thus became an era that called into question both the role and legitimacy of institutions and structures (O'Brien 2006). This increasing scepticism lessens the effectiveness of resilience building.

As society tries to cope with new circumstances, such as climate change, in an increasingly sceptical world, the evidence of risk and an understanding of its consequences must be clear because risk management, in terms of governance, requires the agreement and coordination of many actors. When consequences are easily understood, for instance the dangers associated with automobiles, rules governing use, such as speed restrictions, are usually accepted, if not welcomed. In this instance there is a clear relationship between the different actors in the risk management chain, from the legislators enacting rules to the vehicle driver recognising the reasonableness of the speed restriction. On the other hand, arbitrary or unreasonable rules are often ignored. But in a post-normal world, with the uncertainty associated with climate, the challenge is even greater. In order to make progress, there are some formidable barriers to overcome:

- getting agreement that climate change is a real threat and we need concerted action to deal with it;
- persuading governments to develop flatter, more enabling frameworks that encourage bottom-up efforts;
- find ways to encourage people to believe that climate change is real and that their efforts are needed to build resilient societies;
- overcoming the lack of trust that is fostering scepticism and denial and finding ways to communicate clearly and transparently at all levels and in all directions.

This is a considerable challenge and will only take us to the starting line. There needs to be a new understanding and partnership for resilience that will take a willingness to learn as well as to do things differently. There is a considerable amount that we have learned from disaster management about how people and communities respond to, cope with and recover from disruptive events. This is of great value, but we need to learn how we can prepare, i.e. build resilience prior to events.

Conceptualising resilience and disaster management

The term resilience is usually understood as bouncing back or recovering from a disturbance. This notion implies that resilience is about returning to the original state. But human–environment interactions imply that there cannot be a return to the original state because the nature of the interaction will change the state. Attempts can be made to return the state to its original conditions; however, it is impossible to recreate exactly the same conditions that existed prior to the disturbance. The environmental conditions in which life has flourished are partly a function of the disequilibrium of the system. The change from day to night and from season to season constantly modifies the environmental conditions, and forms of life that have been able to adjust to these changes have flourished. Figure 6.3 shows that, in reaction to a disturbance, the flow of events leads to an often chaotic back-loop, where innovation leads to the development of a new conservation trajectory. This trajectory may be similar to the system that experienced

138 The concept of resilience

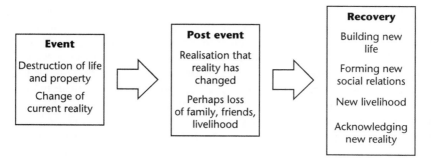

FIGURE 6.8 Resilience as bouncing forward

the disturbance but it will never be identical. The release of energy and resources and the reorganisation will inevitably lead to the development of new trajectories. Prior to the Industrial Revolution, biomass was the most commonly used fuel. The exploitation of fossil fuels led to a new trajectory. Biomass is still used in industrialised countries but mainly for effect or specific uses, for example wood-burning stoves, which do provide heat and some light but are normally an addition to a household that is mainly powered by fossil fuels. Today, fossil fuels dominate the energy system. Despite this domination, the majority of people rely upon traditional biomass.

As discussed earlier, energy poverty is a global problem but despite efforts to address this problem, it is unlikely that all people in the world will be grid connected. Simply put, a top-down technological solution is unlikely to succeed because it leaves people out of the problem. We need to think from a more local people-focused perspective. A disturbance, such as the introduction of renewable systems, following the guidelines established in Chapter 1, would release the energy and resources of these communities and, through a process of innovation and learning in the back-loop, their livelihoods would be transformed. The resilience in this process is the ability to act on the disturbance and use it to deliver long-term benefits. What happens is not a process of bouncing back but moving or bouncing forward.

There is acknowledgement that resilience is a bounce forward ability in the IPPC Special Report on Managing the Risks of Extreme Events and Disasters to Advance Climate Change Adaptation (IPCC 2012b). This offers a definition of resilience that adds 'or improvement' to the UNISDR definition given earlier on page 128. But in the case of extreme events bounce forward ability could lead to transformation as a disaster could redefine individual or household circumstances.

Resilience in disaster management is about ensuring that people are able to respond effectively to disturbances and bounce forward. Figure 6.8 summarises events just after the occurrence of a disaster. In Figure 6.8, it is assumed that at the occurrence of the disaster there are some survivors. They will witness destruction and possibly death. Their initial response will be to protect themselves and then to help others. The reality of their surroundings would be very different from the initial conditions. During the post-event period, a growing realisation of the change in reality would start. The strength of that realisation may be heightened by the loss of friends, family and livelihoods (or some combination of these). During the recovery period, survivors would begin to form new social relationships and either try to rebuild or start a new livelihood. Some may stay in the area and others may move to start a new life in another location. In all cases, the survivors will have moved on, bounced forward or been transformed as a result of the disaster. But transformation can have negative aspects. Some individuals may find it impossible to move on or bounce forward after the loss of loved ones.

Resilience and sustainable development

The term sustainable development was popularised by the Brundtland Commission and is defined as '...development that meets the needs of the present without compromising the ability of future generations to meet their own needs' (Brundtland Commission 1987). Since that time, the term sustainable development has become increasingly embedded within policy discourses (Redclift 2005). There is little doubt that the term has become increasingly used in a variety of fields. Large corporations pursuing a sustainability agenda by producing green products have used the term loosely despite it being described as an oxymoron in terms in economic growth (Daly 1990). On the other hand, green groups use the term to argue for the preservation of an ecosystem.

The contested nature of sustainable development, with theories shaped by people's and organisations' different worldviews, influences how issues are formulated and actions proposed. Diagrammatically, in Figure 6.9, sustainable development is presented as the intersection between environment, society and economy, which are conceived of as separate, although connected, entities and presented as being of equal scale and, presumably, importance. In reality, these are not unified entities; rather they are fractured and multilayered and can be considered at different spatial levels. The market economy is often given priority in policies and the environment is viewed as separate from people. They are, however, interconnected, with the economy dependent on society and both dependent on and located within the environment. The separation of environment, society and economy often leads to a narrow techno-scientific approach, while issues concerning society that are likely to challenge the present socio-economic structure are often marginalised, in particular the sustainability of communities and the maintenance of cultural diversity.

It is possible to elaborate the term sustainable development and establish a number of principles to guide policy making to ensure a holistic approach to sustainable development would be realised. These principles are:

- futurity: regard for the needs of future generations;
- equity: covering social justice regardless of class, gender, race, faith or origin;
- participation: concern that people are able to shape their own futures; and
- environment: respect for biodiversity and ecosystem integrity.

140 The concept of resilience

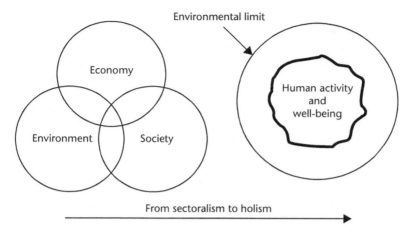

FIGURE 6.9 From sectoralism to holism
Source: Giddings *et al.* (2002).

In reality, policy development is often single-sector focused, for example many activities in the energy sector ignore up- and down-stream impacts. Although holistic methods, such as life cycle assessment, are available, they are often cited as being too costly or time consuming. Checklist or appraisal methods to ensure the sustainability of projects or developments only offer a veneer rather than a solid surface for development. Criticising such inadequate methods is not a rejection of the concept of sustainable development. It is the interpretation and consequent impact on policy development and implementation that is of concern (Hopwood *et al.* 2005).

As discussed in Chapter 1, there are three broad levels of interpretation of sustainable development: status quo, reform and transformation. Within the sustainable development debate the most powerful voices are those of the OECD, essentially the developed world. Status quo, with some minor changes, is the predominant view. Given the power and wealth of the developed world, its influence in shaping global discourses cannot be overlooked (Hopwood *et al.* 2005).

Reform and transformation have powerful advocates. Reform covers a wide range of people but is largely dominated by academics and mainstream NGO experts. Reformists argue that large shifts in policy and lifestyle, many very profound, will be needed at some point but assume that they can be achieved within present social and economic structures, the key being to persuade governments and international organisations, mainly by reasoned argument, to introduce the needed major reforms. Reformists focus on technology; good science and information; modifications to the market; and reform of government, with themes such as dematerialisation, dramatic increases in energy efficiency and a shift to renewables, which are argued to have wider economic and social benefits as well as protect the environment.

Transformation sees mounting problems in the environment and/or society as rooted in fundamental features of society today and how people interrelate and relate with the environment. Those within the sustainable development debate, mainly from environmental justice and indigenous environmental movements, see the fundamental problems as rooted in our present society, which is based on the exploitation of most people and the environment by a small minority of people.

The status quo advocates within the developed world can argue, with some justification, that they have considerable experience and expertise in formulating solutions to environmental problems. This is the case, but the drivers have largely been a response to the consequences of technological development and market standards as opposed to a deep-rooted concern for the environment. The focus of policy development has been risk reduction, usually technologically focused. It is only more recently that policy with a stronger environmental stance has emerged. This shift is largely driven by the increased wealth of the developed world, enabling concern for impacts on well-being (quality of life) to be factored into policy development, a phenomenon where income inequality first increases and then declines with economic growth, postulated by Kuznets in the 1950s and later applied to the environment (Grossman and Krueger 1995). Underpinning the developed world approach to environmental management are strong links to market-based approaches. For example, ecological modernisation originating in the 1980s has evolved from a dematerialisation agenda to one that incorporates social and institutional change but still favours market-based approaches and incremental change as opposed to purposeful interventions. Emerging from this are a suite of measures, collectively termed New Environmental Policy Instruments (NEPIs), aimed at influencing behaviours driven by market-based measures.

A marketised approach clearly shows that economic concerns are dominant and that status quo is the principal interpretation of sustainable development within the developed world. As opposed to holistic and anticipatory approaches to environmental challenges, national frameworks in the developed world continue to evolve in a reactive manner. Given the influence of the developed world model of development, then, arguably, the industrialising nations will follow a similar pattern.

The predominant approach to sustainable development is governed by economic considerations and develops incrementally in response to perceived concerns. Solutions are dominated by technology, often without sufficient recognition of technology as the cause of the problem. This is a weak approach to sustainable development and does not provide an effective international template for development.

Resilience and technology

The concept of resilience is commonly used in technological systems. Resilience in technology focuses on efficiency, constancy and predictability, all attributes at the core of engineers' desires for failsafe design; however, a series of what appear to be rational decisions can increase vulnerability in technological systems despite the intentions of designers to increase system resilience. Much of the infrastructure needed, particularly in cities and the connections between cities, to sustain the socio-economic systems are vulnerable. Utilities, communications, transport systems and industrial and commercial infrastructure are all at risk from climate threats. This is particularly the case in developed world cities, but this problem will continue to grow as the world is becoming increasingly urbanised.

Technological resilience is defined by the UK Cabinet Office as '...the ability of assets, networks and systems to anticipate, absorb, adapt to and/or rapidly recover from a disruptive event' (UK Cabinet Office 2011).

The UK Cabinet Office conceptualises resilience as made up of four components:

1. resistance: built-in strength or protection to hazards;
2. reliability: designed to operate under a number of conditions;
3. redundancy: network design to ensure back-up or spare capacity available;
4. response and recovery: good planning to ensure fast and effective response to and recovery from disruptive events.

This is a very technologically focused approach that does not refer to people – this is hardly surprising because it reflects the weak interpretation of sustainable development by the United Kingdom. Complex infrastructure is largely found in cities and its role is to support the range of cultural, spiritual, administrative, political and socio-economic activities of the inhabitants of the city. In Chapter 1, we discussed the vulnerability of the energy system, a critical infrastructure. By ignoring people in the resilience equation, vulnerability will increase. There is a clear need for more localism in all our systems. O'Brien and Hope (2010: 7552) define a resilient energy system thus:

> A resilient energy system acts to minimize vulnerabilities and exploit beneficial opportunities through socio-technical co-evolution. It is characterized by the knowledge, skills and learning capacity of stakeholders to use indigenous resources to deliver energy services. A resilient energy system exhibits adaptive capacity to enable it to cope with and respond to actual or expected resource changes.

One of the problems with highly interconnected technological systems is their propensity to fail. As Perrow (1999) points out, failure is almost inevitable because errors in either design, construction or operation are very likely, meaning that at some time a failure will occur. For single, isolated systems this may not be much of a problem, except perhaps to the person who is affected by the failure, but today many of our systems are developed in a top-down manner and faults tend to cascade through the system. A clear example of this is the electrical system, where a fault in one part of the system can cause other parts of the system to shut down. In Chapter 1, we argue for a system that has components of both top-down and bottom-up because this is will tend to make it more democratic from a user perspective as well as ameliorating system vulnerability. It will not eliminate system vulnerability, for example the Internet is a flatter and more democratic system but is still prone to attacks from spammers, hackers and viruses as well as technical failures. Users often warn each other of attacks and offer solutions to various problems they have encountered. Of course many businesses have been set up in order to make money by developing and selling solutions, nonetheless there is still a worldwide community that works in a democratic manner to help each other. In terms of technical problems, the architecture of the Internet (bottom-up) allows traffic to be switched to other routes. You may see a slowdown, but rarely do you see a blackout.

This does not mean that technology does not play a positive role. Technology itself is neutral; it is how we use it that counts. The floods in England and Wales during the summer of 2007 were the most expensive in terms of flood damage anywhere in the world during that year. Approximately 50,000 properties were flooded, water supplies to 350,000 homes disrupted and insurance industry costs of GBP 3.2 billion accrued, including GBP 660 million in damage to critical national infrastructure and essential services (Pitt 2008). A UK Government review of the floods, known as the Pitt Review, encouraged the development of a specialised site-specific flood warning service for infrastructure operators (Pitt 2008). During 2009/10, the Environment Agency piloted an Internet-based service, providing targeted flood warnings for infrastructure assets operated by a major electricity distribution company, based on a register of assets identified by geographic x/y coordinates. The Environment Agency subsequently enabled all Category 1 and 2 emergency responders to access a targeted flood warning service from September 2011 (Rhodes 2011).

The development of this pilot scheme demonstrated that a web-based application is one method of delivering a tailored warning service and that 'live flood warning data' held by the Environment Agency could be used to develop a range of specialised products and services. Any organisation can now be licensed to access live flood warning data, retrieving this information from a secure data distribution hub. Whilst the Environment Agency retains a core flood warning service, private sector value added resellers (VARs) are actively developing specialised warning services for utility companies, transport operators, specialist insurance products and supermarket businesses, to name just a few, and have recently taken this further with individual smart phone applications, including a flood warning application for Facebook users (Shoothill 2012). Such services ensure that more people can register for and receive location-specific (targeted) flood warnings in the way they want to. The adoption of these services can contribute to improved flood response, reducing the impact of flooding on communities due to disruption to essential services and critical infrastructure (Rhodes 2011).

The recently introduced 'Facebook flood warning alerts' in the United Kingdom has the huge potential to be adopted globally, wherever suitable baseline mapping and mobile telephony infrastructure exists. The cutting edge of this is its potential in high-risk flood areas, such as East Asia where, using Indonesia as an example, approximately 6 million citizens use Twitter and 42 million use Facebook, many via smart phones (United Nations Office for the Coordination of

Humanitarian Affairs (UNOCHA) 2012). This emphasises the need to recognise that skills and capacities for resilience exist outside typical systems of authority, none more so than in the private sector and communities identified as being at risk.

Resilience and cities

With about half of the global population living in cities, a figure that will increase, making cities resilient is a key challenge. Many cities are in coastal and/or estuary areas making them vulnerable to climate-driven storm surges and long-term sea-level rise. Cities are vulnerable to a range of climate threats. We have seen a large of number floods and storms with considerable impacts on cities in 2012. As extremes are becoming more likely, the probability of similar events will increase. In addition, temperature rises could introduce new disease vectors and coastal erosion, landslides and floods will also be considerable threats.

Cities are centres of cultural, spiritual, economic, learning and communications activities. They are diverse and sometimes chaotic. Resilience for cities has two dimensions: the people and the built environment and infrastructure. UNISDR launched the 'Making Cities Resilient' campaign in 2010 with the ten-point action plan shown in Box 6.2 (UNISDR 2010). This places a strong emphasis on local government to build resilience and on increased support by national governments to cities for the purpose of strengthening local capacities. Local governments and their relationship to sub-national and national governments are varied – there is no single model. Local government functions have evolved, driven by local customs and circumstances, but it should be noted that, typically, local government is involved in both pre-disaster planning and recovery from disruptive events.

BOX 6.2

TEN-POINT CHECKLIST – ESSENTIALS FOR MAKING CITIES RESILIENT

1. Put in place organisation and coordination to understand and reduce disaster risk, based on participation of citizen groups and civil society. Build local alliances. Ensure that all departments understand their role in DRR and preparedness.
2. Assign a budget for DRR and provide incentives for homeowners, low-income families, communities, businesses and public sector to invest in reducing the risks they face.
3. Maintain up-to-date data on hazards and vulnerabilities, prepare risk assessments and use these as the basis for urban development plans and decisions. Ensure that this information and the plans for your city's resilience are readily available to the public and fully discussed with them.
4. Invest in and maintain infrastructure that reduces risk, such as flood drainage, adjusted where needed to cope with climate change.

> 5. Assess the safety of all schools and health facilities and upgrade these as necessary.
> 6. Apply and enforce realistic, risk-compliant building regulations and land use planning principles. Identify safe land for low-income citizens and develop upgrading of informal settlements, wherever feasible.
> 7. Ensure that education programmes and training on DRR are in place in schools and local communities.
> 8. Protect ecosystems and natural buffers to mitigate floods, storm surges and other hazards to which your city may be vulnerable. Adapt to climate change by building on good risk reduction practices.
> 9. Install early warning systems and emergency management capacities in your city and hold regular public preparedness drills.
> 10. After any disaster, ensure that the needs of the survivors are placed at the centre of reconstruction with support for them and their community organisations to design and help implement responses, including rebuilding homes and livelihoods.
>
> Source: UNISDR (2010)

Pre-disaster planning can include negotiation of mutual aid protocols with adjacent jurisdictions and stockpiling equipment and goods likely to be required for response to a hazard event. It can also involve a wide variety of precautionary measures that include hazard assessment, awareness raising, land use and other regulations that reduce building and other uses in exposed spaces and even measures to increase the stability of local people's livelihoods through micro-credit and encouragement of economic diversification. These more elaborate pre-disaster functions are most often carried out in collaboration between local government, sub-national or national government agencies and civil society organisations. But the picture is very varied (O'Brien et al. 2012).

Many populous cities are located near the coast, making them vulnerable to sea-level rises in the longer term. The enhanced social vulnerability created by unplanned urbanisation in many parts of the world will be further amplified by climate change and variability. Many issues will have to be tackled at the level of the city. Little work has been done in terms of vulnerability assessments, city disaster risk assessments or systematic responses for mainstreaming sustainable regulatory frameworks and codes into urban management practice (Parnell et al. 2007).

The core of the public policy challenge for local government is that policy with a focus that remains fixed on promoting institutional resilience and an approach to solving the problems of sustainability that denies or erodes local community entitlements erodes the notion of governance and cannot be regarded as sustainable. Building resilience requires that we need to rethink our approach to disaster management by recognising that resilient responses require a broader engagement and acknowledging the importance of control (being part of the solution as opposed to being dependent), coherence (access to clear information to help understanding and reduce uncertainty) and connectedness (social interconnectivity, being a part of the whole), which builds social capital. Public policy focused on DRR is, at the least, an ongoing task of negotiating entitlements and obligations, and necessarily a social learning process (O'Brien and O'Keefe 2010).

Resilience and climate adaptation

As discussed in Chapter 1, climate negotiations have been very protracted. From a resilience perspective, the failure to agree a timetable and target for emission reductions suggests that the negotiating process is resistant to change, particularly in the developed world. This is similar to the view of sustainable development within the developed world, i.e. to retain the status quo with some incremental change. With no immediate prospect of GHG reductions, hazards driven by climate change are likely to increase. Resilience within hazards research is generally focused on engineered and social systems, and includes pre-event measures to prevent hazard-related damage and losses (preparedness) and post-event strategies to help cope with and minimise disaster impacts.

What is the role of resilience in climate adaptation? We can address this question by recognising the differences between DRR and CCA. Both DRR and CCA are closely linked to development processes. Research on the resilience of social–ecological systems provides some lessons for addressing the gap between these objectives. Yet strengthening the links between DRR, climate change adaptation and sustainable development will not be unproblematic because there are different interpretations of development, different preferences, prioritised values and motivations, different visions for the future, and many trade-offs involved.

The changes in extreme event frequency and intensity associated with climate change can, to some extent, be managed as part of larger efforts to reduce anthropogenic climate change through reductions in GHG emissions. More importantly, however, the drivers of disaster risk can be addressed as a way not only to reduce the losses associated with climate extremes but also as a way of facilitating social and economic welfare and resilience. The challenges posed by climate extremes can provide additional impetus to address existing disaster risks, creating positive outcomes for humans and the environment. A growing literature suggests that a resilient and sustainable future is a choice that involves proactive measures including learning, innovation, transition and transformation.

Summing up

Resilience is a dynamic concept that is used in a number of disciplines. The dynamic nature of the resilience concept fits well with our understanding, for example, of the development of civilisations, their eventual decline and their transformation into new eras. Of course such change takes place over long periods of time, but we can envisage resilience at much shorter timescales, such as the emergence of a new fashion or consumer trend. In addition, resilience as a concept helps us to understand how we can recover from disruptive events and in the process of doing so, move on or bounce forward to a new trajectory. An important aspect of resilience is learning, i.e. drawing from experience and knowledge in ways that are democratic and inclusive. Social learning is an important aspect of resilience building and is discussed later in this book.

But can resilience building be effective in today's world? As we have seen in disaster management, we need to move away from a top-down approach and free ourselves of assumptions, such as 'victims are helpless and need to be helped by professionals'. People need to be part of the solution. Progress is being made but there is still a long way to go. In terms of sustainable development, the current pre-dominant thinking — top-down — militates against effective progress. A top-down approach tends to minimise learning as a strong component of resilience building because there is little two-way traffic. There is consultation but, as the *Devils Dictionary* points out, consultation is just a way of seeking approval for a course of action already decided upon (Bierce 1993). Resilience building in cities, adaptation, technology and sustainable development need a democratic governance system where all stakeholders are engaged. Trying to impose a resilience concept on existing structures implies a view that 'you are responsible for yourselves'. We have no problem with that, providing that resources and power go along with the responsibility.

7
DEVELOPMENT

> No society can surely be flourishing and happy, of which the far greater part of the members are poor and miserable.
> (Adam Smith (1723–1790))

Ideologically, many people seem to be opposed to planning. This is rather wrong because to run a household, a business or a school always requires a level of planning. What good planning needs, however, is that the risks and assumptions behind the plan will continuously be revisited as plan implementation unfolds.

In many ways, national state planning is associated with the 5-year plans of the former Soviet Union. Even before the Second World War, the Indian communist party, long before independence, had established a 10-year plan to implement development. During the Second World War, and for some time afterwards, the British Government had a series of plans built around notions of essential development backed by rationing, a situation that lasted well into the 1950s. The allied victors of the Second World War, in establishing the Bretton Woods institutions, made clear that they were using planning through the International Bank of Reconstruction and Development (IBRD) to rebuild the collapsed economies of Germany and Japan. It was this institution that refocused itself in the 1960s, now known as the World Bank, to deliver development to the newly emerging economies of Asia, Africa and Latin America.

The history of official development policy can be written in four broad chapters. The 1960s emphasised economic growth through modernisation, perhaps iconically captured by Rostow's stages of economic growth (Rostow 1960). It was consciously developed to counteract the stages of development by Marx, suggested in the communist manifesto, from a perspective of capitalism. Instead of Marx's drive from capitalism via socialism to communism, Rostow proposed that capitalism reign supreme, achieving its height when it reached an era of mass consumption. The 1970s saw the consolidation of the decolonisation process started after the Second World War. The new leadership in developing countries wanted to usher in a period of redistribution with growth. This was perhaps best epitomised by the drive and ultimate failure of Tanzania's Ujamaa, which captured the imagination of many intellectuals involved in the development debate, but like many state attempts to drive modernisation, it required a substantive movement of the peasantry, both materially and spatially, and as such destroyed traditional livelihood strategies without delivering the new utopia.

The failure was seen as one that required a substantial change of direction in the 1980s. This was largely led by a rejection of the state as a vehicle for the delivery of economic growth. It partly came about because after the significant increase in oil revenues during the 1970s, developing country governments had borrowed substantially from international markets at what seemed to be cheap interest rates. The over-commitment of central government finances plus an increasing unwillingness of Western capital to commit to development programmes led to a call for structural adjustment by the IBRD's sister institution, the IMF. The essential logic behind this was to lower state borrowing and open up export markets for developing countries' products. The reality, however, was that, with notable exceptions such as Ghana, structural adjustment policies did not seem to work.

The collapse of the former Soviet Union happened in 1989 and brought with it more general talk of open market economies in a globalising world. By this time, the neo-liberal Washington Consensus had emerged, which was basically a commitment to a radical neo-liberalism with little time for the role of the state. It produced two competing ideological interventions. The first was an emphasis by the neo-liberal critics on the lack of governance, by which they meant the lack of Western-type democracies in what they judged to be failed states, such as Myanmar, Somalia, Zimbabwe and Nicaragua; however, many of the characteristics of poor governments they cited, especially the rise of one-party states, were contradicted by the second experience of China lifting 300 million people out of poverty in less than a decade, the single largest poverty alleviation effort in human history.

The opposition voices were largely arguing for a demand side of the equation around Sen's notions of people's entitlements (Sen 1982). These notions of entitlement were broadly framed within a soft version of human rights and were set out as legitimate aspirations for all. The drive to deliver these entitlements was captured by the development of the MDGs with a focus on halving the number of people living on less than USD 1 per day. Two particular MDGs – the right to education, especially primary education for women, and a decrease in morbidity and mortality, especially associated with childbirth – were seen as central to the success of the overall programme. Across all these debates there were two mainstreaming lenses of gender and environment.

Parallel to the debates on economic development and growth was a growing concern about the environmental impact of the whole modernisation process with underlying drivers of population growth and urbanisation. The population growth debate was essentially divided between the developed and the developing world. The developed world largely took the position that population control was a necessary part of successful development (the exception to this position was a rather strange alliance of extreme right-wing liberals and Catholics who basically set an anti-woman agenda). In contrast, the developing countries argued that the only safe contraception was poverty alleviation, not least because children were the single source of wealth to poor people. While the urbanisation debate has significantly grown since the turn of the century, between 1950 and 2000 urbanisation was barely mentioned except to critique the primacy of capital cities because they tended to dominate urban hierarchies. This dominance was often seen as a colonial inheritance because many capital cities were coastal and provided export enclaves for primary commodity production. The debate has recently changed substantially because it is recognised that now over half of the people of the world live in cities, many of which are prone to climate change. Additionally, many cities, especially in Asia, are built in earthquake zones, thus facing additional hazard.

Both the population debate and the urbanisation debate produced new structures in the UN, notably the United Nations Population Fund (UNFPA) and the United Nations Human Settlements Programme (Habitat), with notable conferences in 1994 in Cairo (UNFPA) and 1996 in Istanbul (Habitat).

There was also a general concern over environment. Ironically, this had first developed in the United States, spearheaded by Rachel Carson's *Silent Spring* published in 1962 (Carson 1962). This led to an environmental movement with an emphasis on 'small is beautiful'. From the late 1960s until the early 1970s, the environmental movement was essentially part and parcel of a broader youth movement. It was marked by the publication of *Only one Earth* (Ward and Dubos 1972), written for the United Nations Conference on the Human Environment in Stockholm in 1972, which set up an environmentally focused UN agency, the United Nations Environment Programme (UNEP).

Ten years later, a volume of scientific studies would attempt to estimate UNEP's contribution (see Holgate *et al.* 1982). In the political discussion around UNEP's actual performance, there was realisation that its two successful interventions, namely the Global Environmental Monitoring System (GEMS), which monitored air conditions at altitude, and its Regional Seas Programme had one thing in common – nobody lived there. The conclusion was that it was pointless looking at environments that do not have people. The political background to this 10-year-review was, however, much more contentious. It occurred at the time of the 'second oil crisis', but also at the height of the Reagan–Thatcher press for the neo-liberal world economic order, which, to put it mildly, was not people-focused at all. There was a backlash from the European social democracies who, together with Japan and Canada, decided to establish a commission to look at the relationship between people and environment in more detail. It took 3 years before the commission surfaced. Initially Nyerere, often seen as the father of Ujamaa, was approached. Eventually Gro Brundtland, then Prime Minister of Norway, agreed to head the commission, which reported back in 1987.

In retrospect, the commission's work, especially the work on energy, shows that global warming does not figure significantly in the discussion. The political reality then was that the ozone problem and acid rain dominated the pollution issues. But this was to change, not least as Brundtland presented the reports to the UN General Assembly. The Brundtland report, *Our Common Future* (Brundtland Commission 1987), covered the 'common challenges':

- population and human resources;
- food security: sustaining the potential;
- species and ecosystems: resources for development;
- energy: choices for environment and development;
- industry: producing more with less; and
- the urban challenge.

Most famously, it produced a definition of sustainable development: 'Development that meets the needs of the present without compromising the ability of future generations to meet their own needs' (United Nations 1987). In turn, the General Assembly asked Brundtland for a report back on implementing the *Our Common Future* in 5 years. Brundtland therefore managed to capture the people–development relationship, not just for the present but also for the future.

The UN Conference on Environment and Development (UNCED – commonly known as the Earth Summit) was established and set for Rio de Janeiro in 1992. At Rio, the lawyers took over from the dreamers (for a full discussion of the evolution of the sustainability debate see Middleton *et al.* (1994), Middleton and O'Keefe (2000) and Middleton and O'Keefe (2003)). There were significant treaties around climate change and biodiversity and, as a last-minute response to the demand of the developing countries, a treaty on desertification. There was not a sense of how these treaties would be implemented, and in the following decade a number of consequent agreements establishing targets were made because the policy environment required evidence-based policy formulation structured around impact analysis. It was the start of self-referencing government where relevance, efficiency, effectiveness and sustainability became the buzz questions. The most notable of these agreements was the one around climate change, namely Kyoto (adopted in 1997 in Kyoto, Japan; entered into force in 2005).

Rio generated a further conference 10 years later called the Johannesburg Summit (World Summit on Sustainable Development) in 2002. South Africa was the obvious place to hold it because a new democratic nation with Mandela was a global icon. Unfortunately, for the content of the Johannesburg meeting, the lawyers who had dominated the Rio conference had given way to bureaucrats. The bureaucrats separated north and south but essentially were unable to progress to any serious global resolution on environmental issues. The same occurred 10 years later, in 2012, when, back to Rio at the Rio + 20 United Nations Conference on Sustainable Development, there were only discussions to talk more and a sense that the drive for the environment as a public good had declined substantially, not least because of the failure to generate a post-Kyoto agreement to address climate change. As the century unfolds, we have fine words but cannot find action.

In academia, all of this discussion has given rise to the notion of sustainability science. It is, however, sustainability science predicated on a mature capitalism. It is not a science that questions whether a capital mode of production is compatible with good environmental practice.

Human rights-based approach to development

> A rights-based approach to development describes situations not simply in terms of human needs, or of developmental requirements, but in terms of society's obligations to respond to the inalienable rights of individuals, empowers people to demand justice as a right, not as charity, and gives communities a moral basis from which to claim international assistance when needed.
>
> (Kofi Annan, UN Secretary-General 1998)

The link between international development and human rights is fairly recent. Both concepts developed independently despite the fact that the United Nations Charter (1945) emphasises the promotion of economic and social development as well as the promotion of human rights in its preamble, and the Universal Declaration of Human Rights (1948) incorporates development-related rights ('right to an adequate standard of living') – thus both texts establish a link between the concepts. Both concepts were subsequently politically advanced by the UN, but their independent status was reinforced by the establishment of separate UN programmes/institutions who largely operated independently of each other – the UNDP (an executive board within the UN General Assembly) and the UN human rights related bodies (mainly the Human Rights Council – formerly the United Nations Commission on Human Rights, and more recently the Office of the High Commissioner for Human Rights).

Similar isolation of the concepts was practised in the development community, aid programmes of the governments of developed countries and academia. Human rights were predominantly perceived as a legal 'problem' – an issue of justice – while development, even though it increasingly incorporated issues like sustainability, empowerment, marginalisation, etc. since the 1970s, was still mainly approached from an economic perspective. At its most progressive, economic development was focused on poverty alleviation, never equity.

The connection between development and human rights was not explicitly made until the 1980s/1990s when the Indian economist Amartya Sen initiated a movement away from the conventional notion of development as economic growth towards an emphasis of the human dimension of development. Sen argued that development could not be measured solely by GDP, which did not capture the distribution of wealth/income among a population, but had to incorporate human capabilities and access to resources (Sen 1999). A direct outcome of this argument was the creation of the 'human development index' by the UNDP, which, in addition to income-based measures, includes life expectancy, literacy, education and standard of living of a country – and thus brought the UNDP closer to the notion of human rights.

The adoption by the UN General Assembly of the 'human right to development' was the next logical step in linking development and human rights. It states in its preamble:

> ... development is a comprehensive economic, social, cultural and political process, which aims at the constant improvement of the well-being of the entire population and of all individuals on the basis of their active, free and meaningful participation in development and in the fair distribution of benefits resulting therefrom.
>
> (United Nations General Assembly 1986)

However, the right to development was academically and politically contested and hardly found any expression in the implementation of development – and thus had almost no impact.

In 1993, the UN World Conference on Human Rights in Vienna developed the Vienna Declaration and Programme of Action, which reaffirms the UN Charter and the Universal Declaration of Human Rights and further links democracy, human rights, sustainability and development. It also established the Office of the High Commissioner for Human Rights and reaffirmed the human right to development.

Since the 1990s, the UN developed and promoted a human rights-based approach (HRBA) to development. In 1997, the Secretary General to the UN called mainstream human rights into all work of the United Nations, which led to the establishment of the UN Development Group–Human Rights Mainstreaming Mechanism (UNDG–HRM) in 2009 as part of the 'United Nations-wide endeavour to integrate human rights into development efforts to combat poverty' (UNOHCHR 2006), and the adoption of the UN Statement of Common Understanding on Human Rights-Based Approaches to Development Cooperation and Programming in 2003.

The rights-based approach represents a shift of the development concept from economic growth to poverty alleviation, which, incidentally, is completely in line with the MDGs. The UN asserts that there is a reciprocal relationship between human rights and the MDGs because they

- have common objectives;
- both provide tools for accountability;
- are progressively realised;
- follow similar guiding principles; and
- gender equality is integral to both human rights and the MDGs (UNDG 2007).

The rights-based approach furthermore indicates a transition from the traditional welfare model of development, where rich countries give 'charity' to poor people, to a model based on justice where everybody has a right to an adequate standard of living and development; consequently people become rights holders and actors, while the state attains a legal duty/obligation to provide at least minimum standards of all human rights specified and agreed on, i.e. it becomes the legal duty-bearer. Furthermore, other actors, for example development organisations, NGOs, private sector organisations, have a moral obligation to respect and promote human rights – they are moral duty-bearers.

Thus, in policy terms at least, the link between human rights and development has been explicitly established.

> The promotion of human rights and the fight against poverty lie at the very heart of the United Nations mandate. The two goals are closely connected and mutually reinforcing, as recognised in, among others, the Vienna Declaration and Programme of Action of 1993 and in the Millennium Declaration of 2000.
> (UNOHCHR 2006)

However, practical application of a rights-based approach to development is hampered by problems, particularly by the fact that '... there is no single, universally agreed rights based approach' (Ljungman 2004). The lack of implementable methodologies means that many organisations apply a human rights perspective to development while claiming to use a rights-based approach. A systematic and consistent application of a human rights-based approach would require a paradigm shift.

The environment in extreme – humanitarian assistance

Elsewhere in this book, we detailed the construction of the climate change discourse, but another discourse was opening up, parallel to those on development and environment, which were on disaster management.

The disaster management discourse largely comes out of a geographic paradigm, namely natural hazards research. It is this paradigm that underpins the climate change debate. Global disaster management is the responsibility of the UN system as well as the Red Cross. Its origin is one of humanitarian principles of which the first command is 'do no harm' (a version of the Hippocratic oath) and the second command is 'do some good'. Its principles are essentially derived from combat and are contained in several revised versions of the Geneva Convention of 1949. These principles have been recently codified as the Red Cross principles for humanitarian assistance (Code of Conduct for the International Red Cross and Red Crescent Movement and NGOs in Disaster Relief, 1994),[1] which all agencies, including international NGOs, are meant to adhere to.

After the Second World War and the establishment of the UN, the critical interventions were largely in natural disasters. As drought dominated natural disasters, it was not surprising that the African droughts of the 1970s and 1980s dominated discussions of best practice. There was a parallel discussion in academia that argued for a level of pre-disaster planning associated with three architects who worked separately and together – Cuny, Krimgold and Lewis. They argued from personal experience as architects that pre-disaster planning, critically paying attention to land-use zonation and building codes, could significantly reduce disaster losses (O'Keefe, 1976, personal communication). Of course land-use zonation and building codes do not usually address the dominant hazard form of drought, but they significantly reduce the risk of storm, flood and earthquake damage.

There was a second element to the academic debate which emphasised that is was not the natural hazard event but people's vulnerability that created disaster risk. That vulnerability, at a global level, was seen to be created by conditions of poverty. Poverty itself was viewed as a dynamic of economic growth in which the process of growth could produce both development and underdevelopment. Thus, the development debate, the environment debate and the disaster debate were conjoined with a people focus.

At the end of the 1980s, with the collapse of the Soviet Union, there was a rise of local wars. These wars really predated the Soviet Union collapse in Somalia, Sudan, Mozambique and Angola. But in a bi-polar world of superpowers, the Soviet Union and its allies against the United States and its allies, such local wars were not seen as intervention points for global international intervention. The struggles in Vietnam, Nigeria and Nicaragua, all major humanitarian disasters of the 1980s, were seen to be beyond unified humanitarian help. This changed with the collapse of former Yugoslavia, initially with Bosnia and eventually with Kosovo. Some places healed, such as Mozambique, but there were significant outbreaks of other crises, notably West Africa, Sudan, Somalia and the Democratic Republic of Congo. None of these crises, though serious in themselves, really altered the pattern of disaster intervention, except insofar as the model for natural disaster intervention, on which the Red Cross and UN base their actions, was simply utilised in complex emergencies.

All of this changed with the 9/11 attacks on the United States in 2001. The West invented an Islamic enemy – or an enemy of Islamic terrorists – and very rapidly went from being the hero of Muslim communities in Kosovo to being an evil scourge with invasions in Afghanistan and Iraq. There were also the ambiguous ambitions in Pakistan and ambivalent relations with Indonesia. Non-Western interpretations of disaster intervention see it essentially as being part of a broader new imperialism from developed countries to ensure the status quo, not least because many of these humanitarian interventions occur under the control of the military.

The budget for humanitarian aid has risen dramatically since the collapse of the former Soviet Union, from roughly USD 500 million to USD 24 billion per year. This 48-fold increase in public expenditure has not been repeated in any other sphere and tends to reinforce the assumption that the West is essentially using humanitarian assistance as a substitute for foreign policy when their neo-liberal experiments fail.

The West has been extremely defensive about its position, particularly the INGOs, who need to retain the reputation of 'doing good' to generate private contributions. This all fell apart after the non-intervention by UN military, with the support of the Western international community, in Rwanda. No one still quite knows the number but the estimates remain between 500,000 and 1 million people who were effectively macheted to death in 100 days in 1994. Much soul-searching resulted from this, including the first international evaluation of international aid and several significant changes (Box 7.1), including a shared reorganisation of the UN humanitarian service.

> **BOX 7.1**
>
> **CHANGES IN THE HUMANITARIAN SECTOR IN RESPONSE TO THE RWANDAN CRISIS**
>
> The 1994 Rwandan genocide and the humanitarian response to the crisis motivated an international collaborative evaluation process proposed by the Danish Government's aid agency Danida: Joint Evaluation of Emergency Assistance to Rwanda (JEEAR).
>
> Fifty-two researchers and consultants were employed to deliver five reports:
>
> 1. Study I: Historical Perspective: Some Explanatory Factors
> 2. Study II: Early Warning and Conflict Management
> 3. Study III: Humanitarian Aid and Effects
> 4. Study IV: Rebuilding Post-Genocide Rwanda
> 5. Synthesis Report[2]
>
> The evaluation process was guided by a Steering Committee representing the international aid community, and managed by a Management Group comprising the heads of the evaluation departments of the Swedish aid agency Sida, Norway's Norad, Danida, the UK's Overseas Development Administration (now DFID – Department for International Development) and the US Agency for International Development (USAID). The cost of the evaluation amounted to USD 1.7 million (Borton 2004).
>
> The *Synthesis Report* of the Joint Evaluation, published in March 1996, merged the findings, conclusions and recommendations of the four separate evaluation studies and contained 64 recommendations (Borton 2004). The report highlighted the shortcomings of the humanitarian response and identified weaknesses that had contributed to these shortcomings in the following areas: response capacity, coordination, the monitoring of the effectiveness of overall efforts, the professionalism of some NGOs, and a lack of policy coherence and accountability mechanisms in the sector generally (Active Learning Network for Accountability and Performance (ALNAP) 2004).
>
> **ALNAP**
>
> A direct outcome of the Joint Evaluation was the formation of ALNAP, the Active Learning Network for Accountability and Performance, in 1997. The network

encompasses many key humanitarian organisations and experts from across the sector. Its 73 members include donors, NGOs and INGOs, the Red Cross and Crescent, UN agencies, academic establishments, research institutions and independent consultants.

ALNAP can be seen as a 'collective response by the humanitarian sector' to 'improve humanitarian performance through increased learning and accountability' and by producing 'tools and analysis relevant and accessible to the humanitarian sector as a whole' (ALNAP 2012). ALNAP's mission statement is:

> ALNAP, as a unique sector-wide active learning membership network, is dedicated to improving the quality and accountability of humanitarian action, by sharing lessons, identifying common problems and, where appropriate, building consensus on approaches.
> (ALNAP 2012)

ALNAP's work is organised into initiatives in which members collaborate where mutual interests exist. For example, ALNAP hosted and managed the Tsunami Evaluation Coalition (TEC), which carried out joint evaluations of the humanitarian response to the Asian earthquake and tsunamis of December 2004, the most intensive study since the Rwandan Joint Evaluation.

The Sphere Project

The Sphere Project is a voluntary initiative established in 1997 as a response to the weaknesses in the humanitarian sector identified by JEEAR (Buchanan-Smith 2003). It was initiated by the International Committee of the Red Cross (ICRC) and a group of NGOs to '… improve the quality of humanitarian assistance and the accountability of humanitarian actors to their constituents, donors and affected populations' (The Sphere Project u.d.) by setting minimum standards for the provision of humanitarian assistance in five core areas: water supply and sanitation, nutrition, food aid, shelter and site planning, and health services.

The project is governed by a board of 18 representatives of global networks of humanitarian agencies but is not a membership organisation. Its vision is:

> Sphere works for a world where the right of all people affected by disaster to re-establish their lives and livelihoods is recognised and acted upon in ways that respect their voice and promote their dignity, livelihoods and security.
> (The Sphere Project u.d.)

The main project output, the Sphere Handbook Humanitarian Charter and Minimum Standards in Humanitarian Response, 'is one of the most widely known and internationally recognised sets of common principles and universal minimum standards in life-saving areas of humanitarian response' (The Sphere Project u.d.). The structure of the handbook is depicted in Figure 7.1.

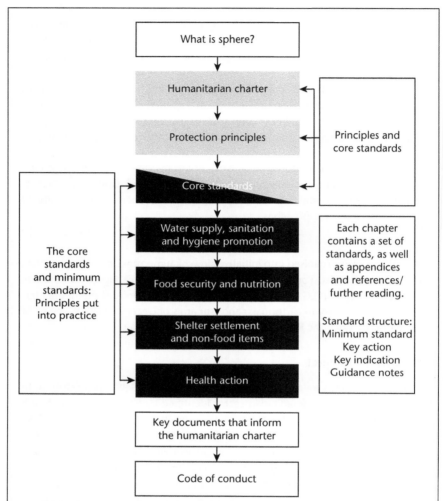

Figure 7.1 Structure of the Sphere Handbook Humanitarian Charter and Minimum Standards in Humanitarian Response
Source: The Sphere Project (2011).

The Sphere Project and Handbook have been criticised by a number of French humanitarian NGOs, including MSF, for being too focused on technical aspects of humanitarian assistance, thereby neglecting both the non-quantifiable aspects and the capacity of the affected population. Furthermore, the separation of technical standards from humanitarian principles in the handbook allowed donors to concentrate on the technical content while 'opting out of the rights based approach' (Kett 2010).

All of these reorganisations, however, insisted that the humanitarian community was self-regulatory; a phrase we have often heard when referring to the banking community in the current financial crisis. The finances of the humanitarian system are relatively straightforward, two-thirds of it are provided by the United States, through the State Department, and by the European Commission. The US State Department regards not just all expenditures but all NGOs who receive money for programmes as an extension of US foreign policy. This somewhat contradicts the notion of humanitarian assistance as being neutral, partial and independent, the so-called Wilsonian doctrine. The EU in contrast has not been allowed to support international organisations, such as the UN, until very recently because individual European member states do not wish the European Commission to have its own foreign policy. The bottom line for humanitarian finance is that little comes from private fundraising. The conclusion must be that the NGOs as well as the international organisations and Western bi-lateral governments are deeply involved in a misappropriation of the international humanitarian agenda for their own purposes.

The UN system and many INGOs usually have a permanent presence on the ground in their countries of operation. They have argued very strongly, especially through the European Commission, that there is a disaster-development cycle that would give them continuity of provision from disaster intervention into development agendas. The unit costs associated with the so-called disaster management cycle (Figure 7.2) are considerably higher than the costs associated with development programmes, ensuring that disaster interventions have higher returns for agencies in capital projects, including transport and IT, as well as higher labour budgets. It is therefore not surprising that the agencies themselves are the biggest supporters of donors being tied to a cycle of funding from disaster to development.

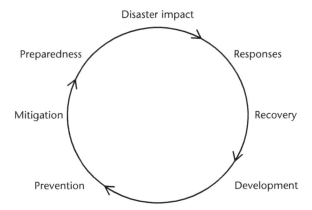

FIGURE 7.2 The disaster management cycle
Source: Carter (1991).

FIGURE 7.3 The disaster–development management cycle
Source: Safran (2003).

The 'traditional' disaster management cycle, as shown in Figure 7.2, sees development as a single phase in the progression through the cycle, whereas a more integrated disaster-development approach (Figure 7.3) perceives prevention and development as synonymous, where development activities would be an integral part of DRR and actively contribute to the resilience and adaptive capacity of the population (see also Chapter 4: Disaster Management).

A subset of this argument is that money spent on DRR in the cycle is the most cost-efficient way of approaching the problem. This argument is being supported by the development of the Hyogo Framework for DRR, which is currently being promoted globally. The expected outcome, strategic goals and priorities for action of the Hyogo Framework for Action (HFA) are clear (Box 7.2) and the economics of it make sense, but the claim that the climate change agenda should be seen as a subset of the DRR agenda is ludicrous. UNFCCC, through Kyoto, has significantly more clout than the UNISDR, through the HFA.

BOX 7.2

THE HYOGO FRAMEWORK FOR ACTION, 2005–2015: BUILDING THE RESILIENCE OF NATIONS AND COMMUNITIES TO DISASTERS

The Hyogo Framework for Action is a global blueprint for disaster risk reduction efforts from 2005 to 2015. It was adopted by 168 governments at the World Conference on Disaster Reduction in Kobe, Hyogo, Japan in 2005. It offers

> guiding principles, priorities for action and practical means for achieving disaster resilience for vulnerable communities:
>
> 'States and other actors participating at the World Conference on Disaster Reduction (hereinafter referred to as 'the Conference') resolve to pursue the following expected outcome for the next 10 years:
>
> *The substantial reduction of disaster losses, in lives and in the social, economic and environmental assets of communities and countries.*
>
> To attain this expected outcome, the Conference resolves to adopt the following strategic goals:
>
> - The more effective integration of disaster risk considerations into sustainable development policies, planning and programming at all levels, with a special emphasis on disaster prevention, mitigation, preparedness and vulnerability reduction.
> - The development and strengthening of institutions, mechanisms and capacities at all levels, in particular at the community level, that can systematically contribute to building resilience to hazards.
> - The systematic incorporation of risk reduction approaches into the design and implementation of emergency preparedness, response and recovery programmes in the reconstruction of affected communities.
>
> The Conference has adopted the following five priorities for action:
>
> 1. Ensure that disaster risk reduction is a national and a local priority with a strong institutional basis for implementation.
> 2. Identify, assess and monitor disaster risks and enhance early warning.
> 3. Use knowledge, innovation and education to build a culture of safety and resilience at all levels.
> 4. Reduce the underlying risk factors.
> 5. Strengthen disaster preparedness for effective response at all levels.
>
> Source: UNISDR (2007).

Most disasters are thought to be site specific: an earthquake limited to a fault line, a landslide to an unstable slope, a flood to a particular drainage regime and a storm to a specific weather system. Climate change is not so site-contained but is a generalised universal problem, which means that it is not a subset of particular hazards. In fact, it is quite the reverse because it is shown to generate obvious meteorological conditions of drought, flood and storm and perhaps even earthquakes (McGuire 2012). It is also associated with the transboundary movement of biological hazards, including changing health ecologies. But the most important reason why the recent DRR agenda cannot subsume the climate change agenda is that the political structures associated with the former are minute while the latter have a mature bureaucracy, strong levels of scientific support and have become the single largest element of environmental negotiation globally. This is not to say that DRR, especially over adaptation, cannot provide useful working examples to the climate change community, but examples do not provide a sufficient basis for institutional change.

The UNFCCC has laboured long and hard to try and fashion an agreement to make meaningful cuts in GHG emissions within a sustainable development context. Developing effective institutions and agreements for global environmental governance is hugely difficult. But things 'are a changing' and perhaps more rapidly than the IPCC Fourth Assessment Report suggested. Recent research strongly suggests that extreme weather events are becoming more frequent and the Arctic amplification may be the culprit in the perturbations in the jet stream (Hansen *et al.* 2012; Francis and Vavrus 2012. This raises the issue of whether or not the current institutional structure of UNFCCC is effective.

Perhaps we need a new global framework – perhaps human and ecological well-being would be a more inclusive way of looking at things. The title is not important, but we do need to focus on the environment (which we need) and people (which the environment does not need). Adaptation is about people and their livelihoods and protecting them from climate risks. This is the role of the UNDP; reducing risk is the role of the UNISDR; and promoting ecological well-being is the role of the UNEP. We could frame all of this with a Convention on Climate Adaptation that could still be linked with the IPCC. We could then streamline the institutional structures and have lines directly to the national, sub-national and local levels. This new body would have sovereign institutional responsibility for developing adaptation within a framework of human and ecological well-being.

There are arguments that focusing all environmental concerns into the climate change debate leads to a carbonisation of the many problems that people have in their interaction with nature. The political reality, however, of global negotiation between the developed and the developing world really does mean that there can only be one war rather than many battles. And even that war over climate change has to be subsumed beneath the collective war on poverty alleviation that underpins the search for an equitable outcome between people themselves and between people and their environment.

8
SOCIAL CAPITAL AND SOCIAL LEARNING

> If you want to have good ideas you must have many ideas. Most of them will be wrong, and what you have to learn is which ones to throw away.
> (Linus Pauling (1901–1994))

Introduction

Social capital, as a concept, has been around for quite some time, but it was not until the 1950s when it emerged in Canadian urban sociology (Woolcock and Narayan 2000) that it began to be used in a number of contexts from urban scholars (Jacobs 1961) to economists (Loury 1977). Broadly, social capital can be built or enhanced through interactions with family members, friends and neighbours. It is from those interactions that we exchange ideas and share knowledge. Building social capital can be thought of as learning in a social context. Both building social capital and learning are important in building adaptive capacity and they are essential processes in resilience building that acts as a counter or antidote to vulnerability. But, as we will see, this building process extends to organisations of all kinds, even to the UNFCCC as discussed later in this chapter.

Thinking about social capital

Extreme events are complex and multifaceted, but research into the impacts of disasters tends to concentrate on the economic impacts and neglect the social aspects, which are more difficult to define and quantify (Tapsell *et al.* 2002; Werrity *et al.* 2007). It has been well documented that many impacts of and responses to extreme events are socially constructed, and that large differences can exist both in the experience of such events and the demands faced by individuals. This social construction is best captured by the concept of vulnerability.

The concept of vulnerability captures such differential impacts by identifying groups particularly at risk. Vulnerability is the outcome of dynamic socio-economic processes and, as such, a complex model, incorporating human, spatial and temporal dimensions (Hilhorst and Bankoff 2004), which is highly variable, area and context specific (Liebmann and Pavanello 2007), and scale-dependent (Vogel and O'Brien 2004). The same can, of course, be said about class but, at base, the definition of class is the distinction between those who work for a living and those who live off capital. The crucial distinction for vulnerability is between those in poverty against those who are materially well off.

Similar to and interacting with vulnerability, the concept of social capital in general and its applicability in policy preparation and formulation in particular has received much attention in recent years from both decision makers and academics (Pelling 1998). Debates on and applications of social capital show that there seem to be more questions than answers available (Field 2008). At one end of the scale, the former UK Prime Minister Tony Blair claimed that social capital is the magic ingredient that makes all the difference in curing all societal ills (Blair 1999). On the other hand, critics claim that 'There is no such thing, other than in the minds of the scholarly careless and/or opportunistic' (Fine 2010: 1).

The social capital concept has not been widely applied to disaster research, and its potential role in disaster preparedness, response and recovery has not been studied much. There are indications that social capital can mediate disaster impacts, certainly on an individual level (Beaudoin 2007; Charuvastra and Cloitre 2008), but its value in the community context is not well understood (Lynch *et al.* 2000; Paldam 2000; Portes 2000).

Social capital and the literature

According to Woolcock (2010), social capital has not merely risen as a social scientific term in the scholarly literature, 'it has become routinized into everyday conversation and policy discourse across an extraordinarily diverse set of disciplines and substantive domains in countries around the world'. Fine (2010: 2) states that the literature on social capital 'has been expanding faster than it can be read and absorbed, let alone be written about'. The popularity of the concept over the last 20 years has led to an abundance of publications (Figure 8.1) of varying quality and with varying objectives. As Woolcock (2010) points out, social capital is an inherently inter-disciplinary concept and, in exploring and reviewing the literature, should be reflected so as not to apply a narrow and limited one-dimensional (one-disciplinary) perspective.

Social capital as a concept

One of the main problems with social capital is that there is no generally accepted definition of the concept. This implies that the measurement of social capital is even less well defined. Li *et al.* (2003) summarise the problems as 'fundamental conceptual and methodological deficiencies', while Bebbington *et al.* (2004) posit that social capital is a 'confused and ill-specified concept based on empirically unsound research'.

Such statements undermine the potential usefulness of social capital as an informative tool for policy making and/or a valuable concept for empirical examination,

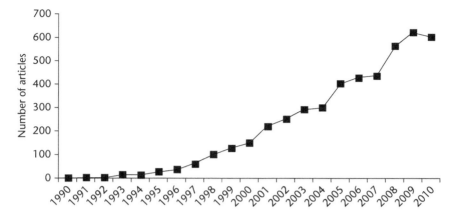

FIGURE 8.1 Frequency of references to social capital in the social science citation index, 1990–2010

Source: Adapted from Halpern (2002) and Field (2008).

but they also indicate the need for further research and clearer conceptualisation of social capital. To dismiss social capital as a useful concept defies its central thesis that 'relationships matter' (Field 2008: 1), which is not only an intrinsically true-sounding statement but one that has been verified by psychological research for many decades (Monnier et al. 1998; Charuvastra and Cloitre 2008; Smith and Christakis 2008).

Woolcock (2010) acknowledges the 'double duty' of social capital as 'providing for diverse audiences a simple and intuitively appealing way of highlighting the intrinsic and instrumental importance of social relationships, while also yielding at the appropriate time to more precise terms appropriate for particular specialist audiences'.

The history of social capital extends much further back than the last two decades of its expanding popularity. Social capital was, however, 'catapulted' into public consciousness by Robert Putnam in 1995, 'social capital's most ardent populariser' (Fine 2002), through his influential – and controversial – article *Bowling Alone* (Halpern 2005: 8). Putnam defined social capital as 'features of social organisation, such as networks, norms and social trust that facilitate coordination and cooperation for mutual benefit' (Putnam 1995). This stretched the use of the concept from an individual asset (as used in psychological research) to an attribute of communities that meant that it could be used to describe a community or even national resource (Johnston and Percy-Smith 2003).

One of the more comprehensive definitions of social capital is given by Farr (2004):

> ... social capital is complexly conceptualized as the network of associations, activities, or relations that bind people together as a community via certain

norms and psychological capacities, notably trust, which are essential for civil society and productive of future collective action or goods, in the manner of other forms of capital.

Social capital refers to the connections between people who have common values, and such networks constitute a resource both for the individual members as well as the group; ultimately 'the concept of social capital boils down to networks, norms, and trust' (Farr 2004) and most definitions refer to these, while also often stressing the aspects of 'mutually beneficial coordinated action' (Evans and Syrett 2007), 'coordinated decision making' (Grootaert and van Bastelaer 2001) and 'information sharing' (Grootaert and van Bastelaer 2001; Hutchinson 2004).

While networks, i.e. the extent and intensity of associational links or activity, are structural aspects, which can be objectively measured, norms and trust largely rely on subjective perceptions and interpretations by individuals. We can distinguish between structural social capital ('what people do') and cognitive social capital ('what people feel') (Harpham et al. 2002).

Szreter and Woolcock (2004) identify three different types of social capital:

1. bonding social capital: ties between like people in similar situations, such as immediate family, close friends and neighbours;
2. bridging social capital: more distant ties of like persons, such as loose friendships and colleagues; and
3. linking social capital: ties to unlike people in dissimilar situations, e.g. people outside the community or formal institutions.

This is an extension of Putnam's earlier distinction between bonding (exclusive) and bridging (inclusive) capital, where bonding constitutes strong ties to like people, while bridging refers to weak ties across groups. The concept of social capital is important in resilience building because this relies on the ties within communities but also the ties with those, not necessarily a part of the community, who are facilitating or working with the group, perhaps as expert advisors. Functioning networks and trust are important aspects of risk management and climate adaptation.

The value of the social capital concept has been recognised beyond the academic arena. The Social Capital Initiative of the World Bank, which began in 1996, produced 24 working papers and the Social Capital Assessment Tool (SCAT) based on a unifying conceptual framework (Figure 8.2), but with the caveat that the tool was to be adapted to the local context in every case (Krishna and Shrader 1999). The approach was based on Putnam's definition of social capital (Putnam 1995).

While conceding the relevance of the macro-level indicators, as shown in Figure 8.2, for understanding social capital in a specific context, measurement is carried out at the micro-level (household, community, and formal and informal institutions) (Krishna and Shrader 1999). The SCAT integrates quantitative and qualitative methods, acknowledging the importance of integrating complementary data collection techniques when trying to analyse a complex concept such as social

Social capital and social learning   **167**

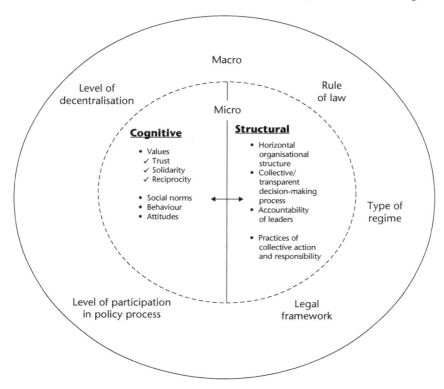

FIGURE 8.2 Conceptual framework for the social capital assessment tool
Source: Krishna and Shrader (1999).

capital. Social capital nestles into the broader debate on governance, which currently drives much developmental thinking. It is also strongly linked to arguments about control at the local level, the subsidiarity debate within the European Union for example.

The Canadian Policy Research Initiative takes a narrower network-based approach to social capital (Figure 8.3) based on their definition: 'Social capital refers to the networks of social relations that may provide individuals and groups with access to resources and supports' (Policy Research Initiative (PRI) 2005).

These different approaches indicate the 'flexibility' of the concept and thus leave it open to debate.

To discuss all the contended issues around social capital would go beyond the scope of this book, thus only a summary of the most prominent debates is presented here:

- whether the term 'capital' should be applied to the concept at all (see, for example, Bankston and Zhou 2002);

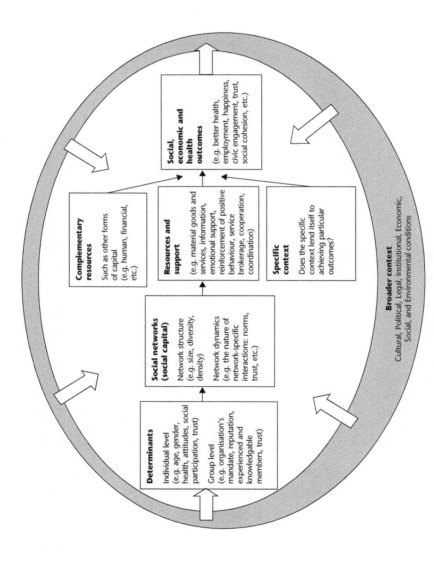

FIGURE 8.3 Conceptual framework for measurement and analysis of social capital of the 'policy research initiative'

Source: Policy Research Initiative (PRI) (2005).

- whether social capital attributes measured at the micro-level can be aggregated to make statements about the quantity and quality of meso- and macro-level social capital (see, for example, Fine 2001; DeFilippis 2002; van Deth 2003);
- the nature of the relationship between cognitive and structural components of social capital and whether the quantity of structural social capital can give any indication of the presence of cognitive social capital;
- whether social capital consists of resources held by individuals or small groups or is a 'process of social interaction leading to constructive outcomes' (Bankston and Zhou 2002) residing in relationships (Woolcock 2010);
- whether the presence of social capital is a prerequisite for, or an outcome of, productive social relations, i.e. 'the direction of causality' (Durlauf 1999);
- how power relations and access to other forms of capital impact on social capital (DeFilippis 2001; Navarro 2002, 2004).

The early implicit assumption about social capital was that it is inherently positive and more social capital will generally be good for individuals and communities (Putnam 1993). It is, however, increasingly acknowledged that social capital can have negative consequences, for example through tightly-knit groups of like people that exclude others or pursue their goals at the cost of others, for example malevolent groups such as the Mafia or when access to networks is unequally distributed and can reinforce inequalities and mistrust (Field 2008). Furthermore, closed communities, i.e. communities with high levels of bonding social capital, can have negative impacts on economic development by hindering expansion and levelling down members' aspirations (Portes and Landolt 1996; Field 2008). Similarly, behavioural norms do not imply positive outcomes; Durlauf (1999) posits that 'social mechanisms which enforce certain types of community behaviour logically lead to correlated behaviours, but do not necessarily lead to socially desirable behaviours', as, for example, in racial segregation. Woolcock (1998) points out that there are different forms of social capital and that the resources these entail should not necessarily be maximised at all cost, but rather optimised. This point parallels the economic conclusions made by Kates (1971) that people's response to hazards, individually and collectively, seeks to satisfice rather than maximise between different economic goals.

Social capital and disasters

It has long been recognised that the social context of disaster-affected individuals and communities is important in determining disaster impacts as well as in mediating the level of coping with such impacts. Despite recognition and acknowledgement of the importance of social relations in disasters, however, research into the detailed interactions between the social context and disaster impacts and outcomes tends to be rather narrow in scope.

If the claims for the effects of social capital are valid, then it should be a valuable tool to enhance understanding of community functioning in crisis, and may have scope for contributing to more effective disaster planning and response as well as building longer-term community resilience.

Social capital has a central role to play in both preparedness and response in the context of a community (Brien 2005), making it an important phenomenon to measure and understand the implications of alongside resilience. There is a basic understanding in the disaster management literature that communities with higher levels of social capital are able to fare better when faced with an emergency event as well as during planning and reconstruction (Murphy 2007). Pelling (1998), however, warns that this is not always the case because social capital can be used to gain power over others. Strong internal links can also lead to a group failing to network to those who are different from them. The real strength of resilience is in the level of inter-community links between various groups (Murphy 2007). Social capital is key to promoting resilience in the context of a community, whether that is in preparing for events or during an emergency event (Wallis et al. 2004).

Few studies relate social capital to the disaster cycle. The ones that have applied the concept to the disaster context observed that social capital may be a prerequisite for adaptive capacity promoting resilience where loose but cooperative community ties were the most effective for a rapid response and recovery (Norris et al. 2008); that social capital increases participation and speed of recovery (Nakagawa and Shaw 2004); that depression was more common in people with low levels of pre-disaster social capital (Beaudoin 2007); and that there is a shift in social capital over the course of the disaster cycle (the increase of collective unity during and shortly after the disaster were replaced by disillusionment and anger in recovery) (Moore et al. 2004). Moore et al. (2004) also observed a re-entrenchment of social inequalities and found that unequal access to post-disaster support and resources were strongly noted by affected people with less developed bridging and linking social capital, notably the poorer and less well educated population. Such findings, however, need to be interpreted with reference to how social capital was actually defined and measured in these studies.

The concepts of vulnerability and resilience are widely used and debated in disaster research with reference to the relatively new concept, or rather desired outcome, of DRR because it is becoming clear now that all are related to – and influenced by – social capital.

Norris et al. (2008) define community resilience as 'a process linking a network of adaptive capacities (resources with dynamic attributes) to adaptation after a disturbance or adversity' and identify four primary sets of adaptive capacities: economic development, social capital, information and communication, and community competence.

It could be argued that adding social capital as yet another ill-defined term to the already complex mix, the 'terminological cacophony' (Bogardi 2006: 2), prevalent in disaster management will only add to the difficulties in operationalisation and further delay advancing and mainstreaming DRR. As discussed, however, social capital is neither a new concept nor has it been absent from conceptual frameworks in disaster research, whether they explicitly state the term or merely imply it.

The development and use of conceptual frameworks in disaster research attempt to simplify and describe the complex interactions at the human–environment interface in the disaster context in order to aid operationalisation of the relevant concepts (Birkmann 2006), primarily vulnerability but increasingly also resilience. As in other areas and disciplines, one of the main problems is to turn academic conceptualisations into policy-relevant and policy-guiding measurable indicators, i.e. to quantify abstract concepts so that they can be 'used' by decision makers. Furthermore, as such operationalisations are developed by separate entities (e.g. government departments, NGOs, etc.) in accordance with their visions and objectives, there are naturally overlaps and gaps in the conceptual and implementation frameworks (Twigg 2006). Cannon *et al.* (2003: 50–59), for example, identify seven widely applied methods of vulnerability analysis and 18 methods under development or less widely used.

Operational frameworks and empirical approaches to the measurement of social capital

The lack of consensus with regard to the constituent components of social capital and its consequent measurement means that most researchers apply different proxies to measure social capital (Johnston and Percy-Smith 2003; van Deth 2003; Voyer and Franke 2006). This is not without problems and Johnston and Percy-Smith (2003) observe:

> ... in order to move the discussion forward it is necessary to return to some very basic definitional questions and develop a methodology that is capable of evaluating the utility and explanatory power of social capital without resort to proxy or surrogate data...

Empirical studies often measure different aspects of social capital (reflected in their chosen indicators/proxies), which are then flagged as a comprehensive assessment of the presence or absence – and sometimes quality – of social capital in a particular context. Apart from relying on a circular argument (social capital is defined as a set of characteristics/indicators whose presence/absence is then taken as evidence that social capital exists in a particular context (Portes 2000), this relies on 'choice variable[s]' (Durlauf 1999) and, as such, may omit other important explanatory variables.

This is fraught with dangers, particularly if policymakers rely on generalisations derived from such studies to inform and justify policy interventions (Woolcock 1998). These varied approaches also restrict consistent interpretation and analysis of social capital data (Green and Fletcher 2003), as well as comparability between findings of different studies.

Despite the lack of a single definition of social capital, van Deth (2003), in reviewing research strategies and empirical approaches, found that concept heterogeneity was not necessarily reflected in operational and empirical heterogeneity.

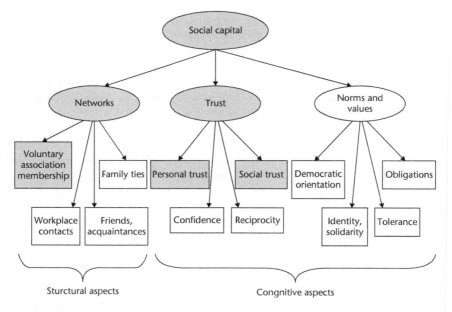

FIGURE 8.4 Most common measurement models of social capital

Adapted from: van Deth (2003).

The most commonly used operationalisations focus on networks and trust, measuring mainly voluntary association membership, personal and social trust, as shown in grey in Figure 8.4.

Measurement of the presence or absence of social capital is, however, not sufficient; social capital as an explanatory variable needs exploration of how it operates within different contexts and on different levels:

> What is required ... is a series of intensive, community-based studies which (as far as is possible) start with a very limited number of hypotheses about the nature, characteristics and consequences of social capital which can then be tested through in-depth, predominantly qualitative, community-based research.
>
> (Johnston and Percy-Smith 2003)

The quantitative tools to measure social capital, such as the SCAT by the World Bank, need further support and validation through qualitative evidence (Harpham et al. 2002; Johnston and Percy-Smith 2003) with a 'more dynamic than static understanding of social capital' (Woolcock 1998).

Some points to keep in mind in contextual investigations of social capital, in addition to the issues discussed above, are that 'ethnic homogeneity is a common feature of circumstances where social capital is strong' (Durlauf 1999); that there is some debate whether the concept can be usefully applied over and above the benefits it yields for individuals or small groups (Lynch *et al.* 2000; Portes 2000; Paldam 2000); and that the definition of the reference area (i.e. community) needs to be validated by local constructs of what community means to people (Harpham *et al.* 2002) rather than imposing geographical boundaries.

This review has shown that the social capital concept is both complex and contested but also that it may be vital in the analysis and understanding of community characteristics and dynamics, and that it is potentially useful in guiding policy development, particularly in a local context. Theoretical discussions in the literature show the need for further conceptualisation and operationalisation of the concept, while empirical studies often lack a firm theoretical underpinning in their approach to the measurement of social capital.

Social capital, like the concepts of vulnerability and resilience, is essentially a fuzzy set of interpretations. Many things are fuzzy in the real world and tightly engineered definitions of these concepts exclude experience. Over the 2012 summer, we have celebrated the London Olympics. From the opening ceremony, which captured how ordinary British people had built social capital – think of the celebration of the National Health Service – to the volunteers who provided social capital to make the Games work, we have seen social capital in action. It is fuzzy, but it works.

Social learning

> Education either functions as an instrument which is used to facilitate integration of the younger generation into the logic of the present system and bring about conformity or it becomes the practice of freedom, the means by which men and women deal critically and creatively with reality and discover how to participate in the transformation of their world.
>
> (Friere 2000)

> Learning would be exceedingly laborious, not to mention hazardous, if people had to rely solely on the effects of their own actions to inform them what to do. Fortunately, most human behavior is learned observationally through modeling: from observing others one forms an idea of how new behaviors are performed, and on later occasions this coded information serves as a guide for action.
>
> (Bandura 1977: 22)

These quotes from Friere and Bandura remind us that learning can be used to reinforce the status quo or it can be a powerful social process that enables people to transform their world. For climate adaptation the use of learning as a transformative process is essential. Social learning and the related concepts of communities of practice and situated learning acknowledge that learning is a social process that occurs in the communities (professional, organisational, place-based) to which we belong. Wenger (2000) posits that through participating in communities of practice, people gain the competences (skills, knowledge and values) that mark them out as members of these communities.

Learning takes place in many ways and forms and in many contexts. Learning is a dynamic and ongoing process. Social learning theory focuses on the learning that occurs within a social context. It considers that people learn from each other, including such concepts as observational learning, imitation and modelling. Social learning is seen as a cognitive process (Bandura 1989; Ormrod 1999; Rotter 1982).

Organisational learning has been part of the management literature for many years and explores how learning takes place in response to changing conditions (Senge 1990; Easterby-Smith *et al.* 1999). The conceptual origins of the learning organisation are closely associated with knowledge management and the increased importance of knowledge as a source of value for companies, institutions and societies, and the advancements in cognitive theory. Organisational learning has its focus on the management of change rather than strategy. Two types of learning are distinguished: single-loop learning on how to do things better and double-loop learning on testing assumptions and rethinking strategies or learning how to learn. Other scholars have developed the concept of triple-loop learning that questions the role of the organisation (Flood and Romm 1996). Learning organisations are able to adapt and respond to change and are seen as being in a state of permanent revolution (Mintzberg *et al.* 1998). These three levels of learning are shown in Figure 8.5.

The science of complexity has generated insights into organisational learning. This parallels natural processes where complex systems innovate and repeat patterns that enhance the ability to adapt successfully to their environment. Complex adaptive systems see groups of autonomous actors that share a goal and have a set of individual and collective rules. The rules may be in tension with each other, and

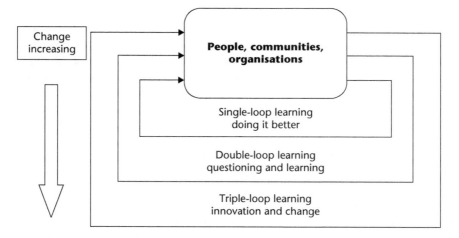

FIGURE 8.5 Learning processes

over time one rule may be replaced by a new rule. This is innovation and leads to change and is seen as learning (Holland 1995; McElroy 2000).

Bowonder et al. (1993) see social learning as a process by which a society or nation perceives, assesses and acts on harmful experiences or past mistakes in purposeful ways. This gives a clear sense that learning seeks a purposeful path and implies a willingness to learn from disruptive events. Great national tragedies, such as the indiscriminate killing of 85 people in Norway in 2011, can have profound impacts on a nation. This type of experience will cause a national evaluation of the ways in which the country should respond. In 1996, the killing of 16 schoolchildren in Dunblane in Scotland by a gunman led to the securitisation of all schools in the United Kingdom.

Keen et al. (2005) see social learning as the collective action and reflection that occurs among individuals and groups as they work to improve a situation. This can be a very powerful process, for example over-the-fence conversations, socialising, community meetings, media etc. and can be very effective in disseminating information and knowledge (O'Brien and Hope 2010).

Both of these perspectives emphasise the importance of social processes for purposeful learning, i.e. learning which occurs as a direct response to disruptive events. People have always been adaptive, but it is the ability to respond in a purposeful way that is important for the integration of climate change adaptation, DRR and development within a sustainable development context. The challenge of climate change means that we will have to use our ingenuity to tackle problems very differently.

In adaptation, social learning is important for transitional or transformational adaptation. This requires a high level of trust, a willingness to take risks, transparency of values, and active engagement of civil society. Learning, of course, is not just about people – it is also about organisations and institutions and how they can change. In the complex geography of bodies that have responsibility for governance for a wide range of activities in which people are involved, it is important that learning occurs at all levels, both horizontal and vertical learning, between and within organisations. There is a need to explore how learning can lead to making smarter decisions about taking a more integrated approach to solving the problems we face.

As climate change adaptation falls within the post-normal arena, effective governance of decision-making is needed. Folke et al. (2005) suggest that one strategy for governing dynamic complex systems in situations of inherent and unavoidable uncertainty is to create governance structures that have capacity for continuous learning and adaptation as new knowledge and new challenges emerge. This fits well with the idea of iteration in both risk management and the development of adaptation strategies, where normal scientific methods continue to research the climate problems we face, while post-normal practice can provide a basis for policymakers.

The role of learning is of vital importance for climate adaptation. There is a real need to manage the conflicting viewpoints between the actors that should be involved in developing climate change adaptation strategies. The key element to successful governance of climate change adaptation is learning.

Organisational learning

Communities and organisations learn, as do people. The climate nexus involves many organisations from global bodies such as UNFCCC through to national governments; scientific communities; academia; international and local NGOs; regional and local governments; state agencies; private sector; and local communities. In short, there are a variety of stakeholders with differing perspectives on and levels of engagement in formulating climate change adaptation options. This raises the issue of how different groups learn, both internally and across organisational boundaries. It also raises the issue of whether or not there has been learning within the climate community and whether this has been effective in developing approaches to climate change adaptation.

Each of the groups in the climate community is a 'community of practice', i.e. it will have its own knowledge base, terminology, standards and code of behaviour. There are many forms of communities of practice throughout society, such as:

- Groups centred on a discipline, such as a research group within a university. Such groups comprise specialist individuals and have a shared interest in a particular field, where their efforts are focused on enhancing their knowledge and understanding in that field.
- Artisan groups that practise a common set of skills. Typically the skills will have been learned through an apprenticeship system where new members work with skilled or experienced people to learn the skills of that craft or trade.
- Groups that are focused on a particular concern or issue, such as an NGO that works to address poverty. Such groups comprise a core staff that raise money and implement programmes to fulfil the mission of the organisation.
- Community groups that have been brought together by a shared concern. Such groups can cover a wide range of issues, such as concerns about a neighbourhood.

A common characteristic of such communities is the learning that takes place within the groups as new information and knowledge is disseminated, discussed and absorbed. New members to the group will also learn as they become familiar with existing members, rules and practices, shared values, the existing state of play and their role.

Members of communities of practice will invest in the group in different ways. Climate scientists, for example, will have invested heavily in developing their expertise, and learning will occur between members as they interact and share new ideas.

New members to an artisan group will bring energy and enthusiasm because they understand that the learning they acquire and the knowledge and skills they will have acquired prior to entry, for example through a school or college education, is likely to ensure a lifetime livelihood.

Community groups will invest their time, local knowledge and their contacts. Some members may bring expertise from either a work setting or a knowledge base. New members are likely to care passionately about the group's concern and will bring enthusiasm and a new perspective. All members will learn from each other as the group dynamics unfold and as they work collectively to address the shared concern.

Those employed by an issue-focused group will bring a skills-set needed by the group and very probably a genuine belief in the mission of the group. These skills could come from their previous employment, their education or life experience, and the introduction of these skills will bring new learning to the group. In addition, the group will learn from its experience as projects are evaluated and lessons learned are fed back to the group.

Organisations are not static and do not work in isolation. For example, it may be necessary for groups that have different perspectives to develop a space or 'boundary organisation' where they can draw on the interests and knowledge of groups on both sides to facilitate evidence-based and socially beneficial policies and programmes (Guston 2001). An early example of a boundary organisation in the climate change debate is the World Meteorological Organisation (WMO) created in 1950. WMO provided a forum for political perspectives to enter into a dialogue with meteorological research.

A more recent example can be envisaged in the interaction between the IPCC (itself a boundary made up of a variety of research disciplines) and member of the Climate Convention (made up of a range of political perspectives and agendas) where space is created for scientists and politicians to explore options. It is at the boundaries between communities of practice where social learning takes place. For this to be effective, Wenger (2000) posits four criteria:

1. an interest or issue that is the focus of shared interaction;
2. a willingness to engage honestly with differences as well as the things held in common;
3. the capacity to see things from the perspective of the other; and
4. the ability to convert existing practice into mutually intelligible forms.

The plethora of communities of practice within the climate change community inevitably means that there is a complex range of alliances that form, change and disband as new knowledge, information, or perspectives are acquired. An example of this could be an alliance between a state and part of the private sector to promote a particular technology. A key challenge is the successful coordination of these alliances and ensuring effective knowledge and skills exchange between the various communities of practice that make up such alliances (Wenger 2004).

Work at the boundary can take a number of forms to promote learning such as knowledge brokers between communities, the shared development of artefacts such as reports and maps, and ensuring effective communications through workshops and websites. Boundary work can enhance social learning through a variety of interactions and interchanges that should lead to a clearer understanding between different perspectives. There is no guarantee that this will ensure harmony and agreement between diverse communities of practice, but the exchange of knowledge, ideas and perspectives does offer the opportunity to develop a shared understanding.

In addition to these more formal methods of social learning, there are other informal ways that shadow formal structures. These informal organisations are networks of friends, acquaintances and contacts that exist externally from communities of practice but also exist with them in shadow form. Such informal networks can offer a space for individuals that is not formally controlled, or sub-groups within organisations to freely experiment, copy, communicate, learn and reflect on their actions (Pelling and High 2005).

Climate change and learning

Both formal and informal networks and the boundary interactions between communities of practice offer considerable scope for social learning. The range of both formal and informal organisations comprising the climate change debate offers considerable opportunities for both horizontal and vertical learning. To what extent has this been the case in terms of climate change adaptation? In an analysis by Nilsson and Swartling (2009), the authors argue that there is evidence that social learning has influenced adaptation policy. The creation of the IPCC in 1988 brought together a variety of experts and was charged with the task of synthesising knowledge to better understand the challenge of climate change. It also brought states into the dialogues as climate change knowledge was seen as an intergovernmental issue. Its first report in 1990 acknowledged that consideration should be given to climate change adaptation if significant adverse climate change should occur (IPCC 1990). But the early days of the IPCC were marked by internal conflict between different interests and a lack of trust between the global north and south. Effectively, little was done to promote the need for adaptation at a political level.

Although initially the IPCC was not formally linked to UNFCCC because of developing countries' objections, the first meeting of COP in 1995 requested UNFCCC's Subsidiary Body for Scientific and Technological Advice (SBSTA) to seek advice from the IPCC on a number of specific topics, including regional impacts of climate change and adaptation responses (IPCC 2004). The reality, however, was that by the time the Climate Convention came into force in 1994, mitigation was dominating its agenda.

Initially, research into climate change focused on impacts and on the development of computer models; however, it became apparent from different bodies of work that some included stakeholders in the assessment and others did not, despite the IPCC's calls for a more integrated approach. The introduction of a vulnerability framework by the IPCC began to assess distributive costs of climate change and the human dimensions of who would suffer the impacts. The inclusion of knowledge from different perspectives in the vulnerability framework, regarding both timeframes and spatial scales, made the new information salient and legitimate to a range of stakeholders (Long Martello and Iles 2006). The Third Assessment Report (IPCC 2001) initiated a flurry of papers on adaptation, resilience and vulnerability, with a considerable increase in papers on adaptation and vulnerability – an indication of the increasing integration of disciplines. Nilsson and Swartling (2009) posit that the IPCC acted to bring together different research traditions, which is an example of how social learning is also relevant for scientific knowledge production, especially in broadening the knowledge base and perspective. This is evidence of learning between communities, but could also reflect the lack of progress on mitigation. Between 2001 and 2007, IPCC assessments report a clear consensus that anthropogenic climate change had emerged. Adaptation became part of the political discourse on sustainable development and was increasingly linked with other international policy initiatives, including the MDGs. The shifting dynamics within the climate change debate saw trans-state networks of indigenous groups gain more prominence, advocating a more human face to the problems through an emphasis on vulnerability.

The analysis by Nilsson and Swartling (2009) shows that adaptation has gained credence at the political level, and this is partly due to a social learning process within and between organisations involved in the climate change debate. Learning has been both horizontal and vertical, the latter evidenced by the growth of national, sub-national and local efforts. But has the growing realisation that efforts (to date) on mitigation that have largely failed contributed to this? Does this indicate that another learning process is present, a process that recognises the futility of mitigation negotiations to date? If so, we term this 'learning of the inevitable', where gradually, as a position becomes hopeless, there are efforts to promote other options that have some hope, even if at an earlier time they were derided, not necessarily because of belief but for reasons of political expediency. This raises the question whether or not this shift in thinking can be translated into effective action.

Learning for resilience, adaptive capacity and risk management

Climate change adaptation strategies aim to build adaptive capacity. This enhances resilience and the ability to manage risks. Social learning has a key role to play in building adaptive capacity and hence building resilience. Adaptive capacity is the ability or potential of a natural or human system to respond successfully to climate variability and change, and includes adjustments in both behaviour and resources and technologies. This includes the ability to moderate damages, take advantage of opportunities and cope with consequences. This is a considerable challenge. Drawing from the disaster management field, it is possible to see that learning can play an important role in resilience building.

Disaster risk reduction and social learning

Building resilience recognises that human–environment interactions are both the cause of increased risk, but also the space where interventions can refocus efforts on preparedness, built on local knowledge informed by predictions of likely climate impacts. It also requires a shift in the behaviour of response agencies, with a greater focus on preparedness achieved through resilience building that enhances coping capacity. The conventional disaster management cycle tends to be locked into a process of single-loop learning where the emphasis is on institutional resilience and improved performance within given parameters. This lacks the richness of local knowledge and experience necessary to build effective preparedness in responding to produced unknowns. We argue that a paradigm shift is needed. The new paradigm needs to focus on people and their environments and use learning methods to develop preparedness. This does not exclude response bodies, which are essential for resilience building, but argues that there are two distinct roles for these bodies. This is shown in Figure 8.6.

The double-loop approach set out in Figure 8.6 recognises the importance of conventional disaster management approaches to routine emergencies. We cannot

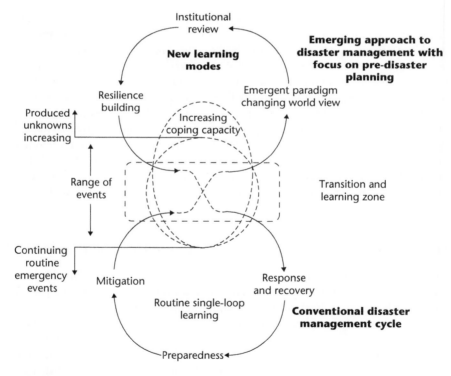

FIGURE 8.6 New learning cycle

Source: O'Brien *et al.* (2010).

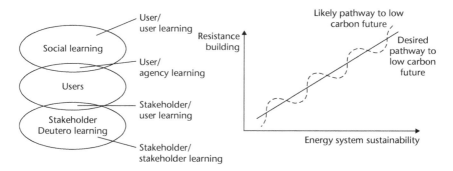

FIGURE 8.7 Learning for a low carbon pathway

Source: O'Brien and Hope (2010).

eliminate risk; however, the second loop ('new learning modes') is where resilience building is approached by doing things differently. The key shift is in the 'transition and learning zone'. Learning from events, we argue, should be a catalyst for thinking about how we do things.

Learning in this way can be transposed to all sectors. O'Brien and Hope (2010) show how learning processes can be used to stimulate the transition to a low carbon pathway through the interaction between different user groups, agencies and stakeholders (Figure 8.7). Essentially, in this process the stakeholders learn that they need to learn (deutero learning) and, through agencies and user groups, identify needs and what works and why. This is a process of iteration that leads to resilience and closely mirrors the iterative approach to risk management and adaptation discussed earlier.

Summary

If we want to build resilience, building social capital through social learning is needed. Simply put, we cannot hope to make progress if there is a lack of willingness at the institutional level to break down barriers. This means accepting that climate change is real with the rider that no one is to blame and we all are to blame. This is where we learn that we have to learn to do things differently. Empowering people and recognising entitlements are the starting point. Social learning is the vehicle for building social capital for resilience building.

9
CONCLUSION

> The times, they are a changing
> (Bob Dylan (1941–present))

Introduction

Readers are probably aware that the paths of the authors have crossed and come together at several points over the last 30 years. A shared togetherness is a politics of equity of opportunity that basically has at its core a commitment to poverty alleviation at local, national and international levels. The successful delivery of poverty alleviation requires a commitment to building sustainable livelihoods, something that we have addressed throughout our 30 years of work. This core is informed by a shared background in science and technology, particularly energy technologies; this sharing obviously brings us to the problems that these technologies have caused, especially the problem of climate change. Climate change itself is a wicked problem where the question seems simple but the answers are multiple, complex and contradictory. The core argument for us about climate change is not simply the change in temperature and precipitation but the increase in extreme events. What we are talking about is increased storm, flood and drought episodes, so-called natural hazards, which can only really be addressed if we understand the vulnerability of the people affected by the event. Understanding vulnerability encouraged us to believe that it is possible to embark on DRR strategies, which in turn encourages detailed exploration of both adaptation and mitigation projects. Core to DRR is the exploration of people's resilience.

The outcome of Rio + 20 clearly showed that the road to hell can be paved with good intentions. Essentially, Rio + 20 has marked a watering down of commitment to sustainable development and thus to resilience. Sustainability has been embedded across the institutional landscape. But two critical questions remain: what is the paradigm of sustainability and what is the level of institutional movement towards greater equity for people and environment? Resilience is the new paradigm for institutional adoption, which again should generate institutional movement. What we have essentially argued in this book is that it is only the radical paradigm of a political economy of sustainability, by implication resilience, that will really work. Traditional ecological approaches in both environmental and economic science, with their emphasis on equilibrium, contain a bias against change. We reject that bias because change to equity must be driven by people, for people. It requires change not equilibrium.

So where next?

As we have pointed out in our criticisms of both the sustainability and resilience paradigms, there is a preference for both physical and social science to base their modelling on notions of equilibrium. Equilibrium points are generated through process through time following a probabilistic or stochastic path. Eventually, equilibrium appears as the stationary point, the balance, of the whole system. Systems, however, become disturbed for external reasons or because of internal collapse and there are sudden shifts, beloved by chaos theory, which move the system onto a different plane. Natural history, including that of evolution, and social history, including the destruction of feudalism, are two examples. These relatively sudden shifts are known as punctuated equilibria (Gould and Eldredge 1977). It is precisely such a punctuated equilibrium that is necessary to resolve the current contradictions of human beings in nature.

For this reason, we have drawn up a table to capture the broad schools of resilience (Table 9.1).

Again, we have three schools: ecology, conventional economics and political economy. Only political economy is people-focused with an emphasis on evolution that empowers people. For this reason we refer to it as bounce-forward ability.

Facing the challenge of building resilience as bounce-forward ability, we need a road map that can generate a point of punctuated equilibrium. That punctuated equilibrium must be a positive social movement away from the form of capitalism that has currently brought international relations to the brink of disaster. A road map exists, courtesy of Harvey's reading of Marx (Harvey 2008). In his

TABLE 9.1 Schools of resilience

Subject matter	Ecological resilience	Conventional economics	Political economy
Major concern	Ecosystems	Market	Livelihoods
Major goals	Ecological viability	Growth	Social justice
Major academic base	Biology, ecology	Finance	Political economy
World view	Equilibrium	Equilibrium	Evolution
Research approach	Neutral	Neutral	Explicit values
Major advantages	Diversity, ecosystem services	Profitability focus	Equity
Major flaws	Not evolutionary	Not developmental	Little formal theory
Policy prescription	Protect nature	Strengthen market	Empower people
Policy delivery	Top-down	Top-down	Bottom-up
Policy presentation	Bounce-back ability	Bounce-back ability	Bounce-forward ability

reading, Harvey indicates there are six moments in the process of human evolution. No single moment prevails over any other and each moment has the possibility of autonomous development. Harvey likens the totality to ecology and such a metaphor minimises the possibility of determinism. The moments are technology, nature, production, reproduction, social relations and mental conceptions.

If we look to technology, we see the challenge in the reduction of climate risk to be effective adaptation. In the long run, mitigation itself is an adaptation strategy. We need a more energy-efficient future where the same level of service is provided for substantially less energy input. We also need a significant move away from carbon-based fuels to a renewable future. This has implications, not just for electricity generation but the whole range of transformation production, transmission, distribution and final end-use. In short, we need to move to a world that is more locally focused. We will never be wireless, as in the recent telecommunications revolution, but we certainly need to be 'wire less'.

If we look to nature, our relationship needs to change substantially. This is not just an issue of carbon emissions and/or ecological footprint, but a challenge to abandon the conceit of capitalism that is the production of nature (Smith and O'Keefe 1980, 1985). Capitalism in producing nature pretends it can master nature as if the laws of capital accumulation were stronger than the laws of thermodynamics; yet, as the current financial crisis has shown, the gross assumption that capital can control nature through an expanding commodification and commoditisation of the global commons is generally impossible. Rethinking human relationships with nature is essential because we cannot materially separate ourselves from our own habitat.

We then need to look at the actual moment of production. Production is dominated by material growth and essentially valorised through the market. There are tendencies in the system to increase valorisation by overproduction, which in turn undermines accumulation strategies. There are also tendencies in the system to design for built-in obsolescence, thus requiring repeat purchases while at the same time generating large amounts of waste material. While the market itself can be a useful guide for best-fit production, it simultaneously hides the fact that it generates inequalities between classes and countries. Production is a social process and the market must be controlled as a social process rather than controlling the social process.

There are then the moments of the production and reproduction of daily life. It seems obscene in the twenty-first century that 2 billion people still live on less than one dollar a day. There is a requirement to consider what is needed at household level because the household is the dominant form of organisation for material exchange to generate an adequate 'basket of income'. Clearly that basket of income must include enough subsistence for the worker to survive. This element of subsistence is significantly less than the elements required for the social reproduction of people: think about education and health costs, childcare and care for the aged – these are real costs that individuals pay for either as private goods through individual purchase or as public goods through taxation. Taking only a simple comparison of

US and UK healthcare systems, the public good of the UK National Health Service offers more efficient healthcare delivery in a more democratic manner than the privately based US health model. Two other elements of the basket of income are important. The first is the cost of environmental reproduction and the replacement value of technologies – even if it is just gardening with a spade, there are significant costs to maintaining environmental productivity. The second cost is the historical and moral gains that people have made individually and collectively, such as rights to health and safety clothing or to additional payment for out-of-hours employment. But when the whole basket of income is looked at, the heaviest cost falls on our own social reproduction as human beings. Capitalism tries to (but cannot) have it both ways, demanding labour but insisting it does not have to pay for the reproduction of labour.

Issues of reproduction and social reproduction are paralleled by another moment, which is the actual social relations between groups. Obviously, under capitalism the first issue is one of class. We like to think about this not in simplistic terms of class antagonism but in the context of how each of us plays out our own lives, the crucial issue here being control of self or time. From our university base in Northumbria, we look at the urban and rural hierarchy to find that, at present, we are run by a government of public school boys whose inherited wealth gives them a position of power. With this power they reinforce the position of the rich against the poor. Relations can change positively, as demonstrated by China where 300 million have been lifted out of poverty in a decade and by the emergence of the other BRIC (Brazil, Russia, India and China) countries. But the power of capital accumulation is one where the power of the capitalist class is reinforced, even if middle classes are allowed to expand. Class relations are not the only issue, however. The inequalities of gender, particularly responsibilities for caring and the social reproduction of the household, are still rife. This tension, together with the ones generated by age, disability and religion needs to be addressed if the moment is to be overcome.

Finally, there is the importance of the production of ideas, a mental process. Contrary to many myths, ideas do not fall from the sky. More importantly, ideas are rarely the work of an individual genius. We need to recognise that most research is produced in public institutions, including the social research of the state itself, rather that private enterprise. We must recognise too that ideas are generated by our material interaction with nature through technology. We must be in that triple-loop learning where resilience is building back better to go forward.

Moving the moments would create contradictions, contingencies and even autonomous possibilities. There will be much unevenness as the moments unfold, but the road map takes us to where we need to be – to an international goal of equal opportunity rather than planning for an equal outcome.

Tipping points

In our post-normal world, laden with uncertainties, we reflect on what we have learned from science. A tipping point in climate science represents a point where the climate system flips from one condition to another. As the climate system warms, we see increased variability and extremes. Hansen *et al.* (2012) have mapped this process. In a nutshell, a normal distribution around a mean global temperature would produce a bell-shaped curve and we would expect to see variation in temperature events, both lower and higher events. But as the mean global temperature rises, the bell-shaped curve would tend to flatten, meaning that the probability of extremes would rise. The road map we referred to earlier may not be possible and our journey may take us to another place.

We have referred to punctuated equilibrium and there are many examples we can think of. The Industrial Revolution drove massive social and technological change. The Communist Revolution in Russia overthrew the Czars and established the Soviet Union. The collapse of the Soviet Union has seen the emergence of a new Russia and a change of relations within Europe. In China, the Communist Revolution transformed the country and the Cultural Revolution almost destroyed it. The reaction has been the emergence of state capitalism that has propelled China to be the second largest economy in the world. In the United Kingdom, huge social change was driven by the end of the Second World War when voters wanted real change. The one thing all of these events have in common is that they affected either a country or a group of countries. Climate change will affect the whole planet. This will truly be a global punctuation.

Since the Industrial Revolution, we have tried to master nature and in the process have systematically pillaged the planet. The drivers of population, lifestyle and a misplaced faith in neo-liberalism, discussed earlier in this book, have taken us to a point where, as Lovelock's Gaia Theory suggests, the global ecosystem is adjusting to the changes (Lovelock 2006). In terms of climate adaptation, this is something we also have to do. We need to use iterative approaches, a sort of trial and error, until we can begin to have confidence that we are beginning to develop effective adaptation strategies. If we do reach a tipping point, then we will need a transformative approach. It is impossible to predict what will happen. But, as with sustainable development, we can map where we want to go with resilience – the status quo, reform or transformation. So far, our institutions are status quo or, at best, reformist not bouncing forward. Some indigenous institutions show an evolutionary design but are hard pressed, usually because of the encroachment of globalisation.

It is reasonable, however, to ask the question 'what went wrong and why?' and we are sure that, when we are propelled into a new and a more uncertain future, many will. The key theme in this book is that we seem unable to work together; governments cannot agree to lower GHG emissions, reduce ecological footprints or tackle poverty. This is political failure. The institutions charged with the responsibility to find solutions have, so far, failed to do so. This is institutional failure. The economic system has thrown the global economy into chaos. This is economic and fiduciary failure. We suppose we should blame ourselves and we are sure that many do, but we have to be part of the solution. Globally, there has been fragmentation in the way we have tried to develop solutions. This suggests that there is a need for reform, both in the way we are governed and the way institutions function.

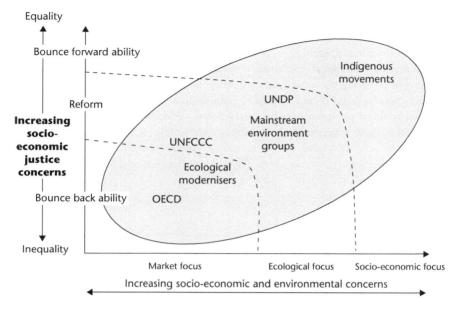

FIGURE 9.1 Mapping the resilience debate

Changing governments seems unlikely. Perhaps we should focus our attention on the global institutions, such as UNFCCC, UNISDR and UNDP, and ask if they are functioning in a way that is likely to produce effective results. Perhaps we need post-normal institutions to deal with post-normal problems.

We offer no solutions but argue that the current drive for resilience, whilst laudable, is an abnegation of political responsibility without the devolved power and resources. If governments want us to be resilient, then they must give us the means and the tools. The direction of departure is from a base of thinking about resilience in a social science context, where resilience may be more popularly defined as the individual's and community's ability to bounce forward after extreme events. Modern everyday life requires that we plan for an unknown future knowing that, at some time, it will happen. The drive must be to put the price to be paid for individual and community resilience into the planning process that matches the way the 'Polluter Pays Principle' has been incorporated into environmental management. This, however, is no easy task because it is not simply an economic or financial consideration but a question of how we seek our entitlements to govern ourselves, including the governance of our own physical and social reproduction.

The real solution lies in addressing issues of governance. Central to the approach to governance must be the emphasis on a system that reduces social risk both in the present and in the future. Such a resilient approach has, at its core, the view that any production of nature or of society should be done from a risk reduction perspective. From such a perspective, adaptation is not an ex-post response to actual extreme events but is normalised in the planning process as a way of approaching known unknowns. Adaptation structure requires both top-down and bottom-up approaches, which may be regarded, respectively, as the deductive and inductive methods of resilient planning.

Solving the ecological and market challenge of sustainability is not progress either, if in resolving these problems the solution denies or erodes local community entitlements. In short, there is a need to both think and learn differently if we wish to build resilient communities and establish a critical but sustainable future. But be careful because when we hear those with power talk of resilience, it is because they do not wish to resource a future. They are saying, 'You the people are by yourselves'. We are not by ourselves. We build resilience going forward together.

We cannot risk our children's children.

NOTES

2 The climate journey
1 Decisions that meet the objective of the Convention can be adopted even if there are still some issues that need agreement, usually postponed to a later date to allow parties to resolve the differences.
2 REDD+ is more holistic than the REDD programme and includes the role of conservation, sustainable management of forests and enhancement of forest carbon stocks.

3 Climate extremes: does a post-normal approach make sense?
1 http://www.opensecrets.org/industries/indus.php?Ind=E

4 Disaster management
1 Often terminology can be confusing, for example the terms disaster management and emergency management. The former implies a more multi-level approach whilst the second implies a more militaristic 'command-and-control' approach as can be seen from the following definitions.
UNISDR does not have a definition for disaster management but defines disaster risk management as:

> The systematic process of using administrative directives, organizations, and operational skills and capacities to implement strategies, policies and improved coping capacities in order to lessen the adverse impacts of hazards and the possibility of disaster.
> (UNISDR 2009)

UNISDR defines a disaster as:

> A serious disruption of the functioning of a community or a society involving widespread human, material, economic or environmental losses and impacts, which exceeds the ability of the affected community or society to cope using its own resources.
> (UNISDR 2009)

UNISDR defines emergency management as:

> The organization and management of resources and responsibilities for addressing all aspects of emergencies, in particular preparedness, response and initial recovery steps.
> (UNISDR 2009)

Effective emergency action can avoid the escalation of an event into a disaster. The terms disaster and emergency management are often used interchangeably. For purposes of consistency the term disaster management will be used throughout this book.

5 Adaptation

1 Further details are available at http://www.nlcap.net/

7 Development

1 The Code of Conduct is available from the IFRC website http://www.ifrc.org/Docs/idrl/I259EN.pdf

   1. The humanitarian imperative comes first.
   2. Aid is given regardless of the race, creed or nationality of the recipients and without adverse distinction of any kind. Aid priorities are calculated on the basis of need alone.
   3. Aid will not be used to further a particular political or religious standpoint.
   4. We shall endeavour not to be used as an instrument of government foreign policy.
   5. We shall respect culture and custom.
   6. We shall attempt to build disaster response on local capacities.
   7. Ways shall be found to involve program beneficiaries in the management of relief aid.
   8. Relief aid must strive to reduce vulnerabilities to future disaster as well as meeting basic needs.
   9. We hold ourselves accountable to both those we seek to assist and those from whom we accept resources.
   10. In our information, publicity and advertising activities, we shall recognise disaster victims as dignified human beings, not hopeless objects.

2 Synthesis Report. *The international response to conflict and genocide: lessons from the Rwanda experience.* Available from: http://www.oecd.org/derec/50189495.pdf [Accessed 15 August 2012].

REFERENCES

Abrahams, J., 2010. Monckton takes scientist to brink of madness at climate change talk, *The Guardian*, 3 June 2010. Available from: http://www.guardian.co.uk/environment/blog/2010/jun/03/monckton-us-climate-change-talk-denial [Accessed 5 June 2010].

Active Learning Network for Accountability and Performance in Humanitarian Action (ALNAP), 2004. *Assessment of the impact and influence of the Joint Evaluation of emergency assistance to Rwanda*. Available from: http://www.alnap.org/resource/3354.aspx [Accessed 15 August 2012].

Active Learning Network for Accountability and Performance in Humanitarian Action (ALNAP), 2012. *About ALNAP*. Available from: http://www.alnap.org/about.aspx [Accessed 15 August 2012].

Adam, D., 2010. How has 'Climategate' affected the battle against climate change? The Muir Russell report has cleared the scientists of any dishonesty over data, but how did the scandal affect Copenhagen? *The Guardian*, 8 July 2010. Available from: http://www.guardian.co.uk/environment/cif-green/2010/jul/08/hacked-climate-science-emails-climate-change [Accessed 15 August 2010].

Akhand, M. H., 2003. Disaster management and cyclone warning systems in Bangladesh. *In*: J. Zschau and A. N. Kuppers, eds. *Early warning systems for natural disaster reduction*. Berlin: Springer, 49–64.

Alexander, D., 1993. *Natural disasters*. London: UCL Press.

Alexander, D., 2002. From civil defence to civil protection – and back again. *Disaster Prevention and Management*, 11 (3), 209–213.

Anderson, K. and Bows, A., 2008. Reframing the climate change challenge in light of post-2000 emission trends. *Philosophical Transactions of the Royal Society* [online], 366, 3863–3882. Available from: http://rsta.royalsocietypublishing.org/content/366/1882/3863.full.pdf+html [Accessed 21 August 2012].

Anderson, K. and Bows, A., 2011. Beyond 'dangerous' climate change: emission scenarios for a new world. *Philosophical Transactions of the Royal Society A*, 369, 20–44.

Ashley, C. and Carney D., 1999. *Sustainable livelihoods: lessons from early experience*. London: DFID.

Asian Disaster Reduction Center (ADRC) and International Recovery Platform (IFRP), 2011. Earthquake and tsunami in Japan's Tohoku region. Rapid damage assessment and need survey. March 21–24. Available from: http://www.adrc.asia/documents/disaster_info/20113.11_Earthquake&Tsunami_in_Japan.pdf [Accessed 2 September 2012].

Atteridge, A., 2011. *Will private finance support climate change adaptation in developing countries? Historical patterns as a window on future private-sector climate finance.* Stockholm Environment Institute, Working Paper No 2011-05. Available from: http://www.sei-international.org/mediamanager/documents/Publications/Climate/SEI-WP-2011-05-Private-Sector-Adaptation-Finance-ES.pdf [Accessed 3 September 2012].

Bandura, A., 1977. *Social learning theory*. Englewood Cliffs, NJ: Prentice Hall.

Bandura, A., 1989. Social cognitive theory. *In*: R. Vasta, ed. *Annals of child development. Six theories of child development*. Greenwich, CT: JAI Press, 1–60.

Bankston, C. L. and Zhou, M., 2002. Social capital as process: the meanings and problems of a theoretical metaphor. *Sociological Inquiry*, 72 (2), 285–317.

BBC, 2010. Climate change talks 'backslide' at Bonn, *BBC News*, 7 August 2010. Available from: http://www.bbc.co.uk/news/science-environment-10900798 [Accessed 9 August 2012].

Beament, E., 2010. Global warming deal hopes revived after Cancún agreement, *The Independent*, 11 December 2010. Available from: http://www.independent.co.uk/environment/climate-change/global-warming-deal-hopes-revived-after-Cancún-agreement-2157688.html [Accessed 3 September 2012].

Beaudoin, C. E., 2007. News, social capital and health in the context of Katrina. *Journal of Health Care for the Poor and Underserved*, 18, 418–30.

Bebbington, A., Guggenheim, S., Olson, E., and Woolcock, M., 2004. Exploring social capital debates at the World Bank. *Journal of Development Studies*, 40 (5), 33–64.

Beck, U., 1992. *Risk society: towards a new modernity*. London: Sage.

Beck, U., 1995. *Ecological politics in an age of risk*. Cambridge, UK: Polity Press.

Beck, U., 2009. *World at risk*. Cambridge, UK: Polity Press.

Berginnis, C., 2005. House reorganization of authorizing committees involving FEMA programs. Re: Letter to the Honorable Dennis Hastert, Speaker of the United States House of Representatives, dated 6 January 2005. Available from: http://www.floods.org/PDF/ASFPM_Letter_FEMAinDHS_House_Jan05.pdf [Accessed 7 September 2012].

Berkes, F. and Folke, C. eds., 1998. *Linking social and ecological systems: management practices and social mechanisms for building resilience*. Cambridge, UK: Cambridge University Press.

Berkley Earth Temperature Group, 2011. Available from: http://berkeleyearth.org/index.php [Accessed 10 September 2012].

Bierce, A., 1993. *The Devil's dictionary, 1881–1906*. UK: Dover Thrift Publications.

Birkmann, J., 2006. Measuring vulnerability to promote disaster-resilient societies: Conceptual Frameworks and definitions. *In*: J. Birkmann, ed. *Measuring vulnerability to natural hazards*. Tokyo: United Nations University Press, pp. 9–54.

Black, R., 2012. Arctic melt releasing ancient methane, *BBC News online*, 20 May 2012. Available from: http://www.bbc.co.uk/news/science-environment-18120093 [Accessed 22 May 2012].

Blair, A., 1999. Tony Blair's keynote speech, NCVO Annual Conference.

Boardman, B., 2010. *Fixing fuel poverty*. London: Earthscan.

Bogardi, J. J., 2006. Introduction. *In*: J. Birkmann, ed. *Measuring vulnerability to natural hazards*. Tokyo: United Nations University Press, pp. 1–6.

Bohle, H. G., 2002. Editorial: the geography of vulnerable food systems. *Die Erde*, 133 (4), 341–344.
Borton, J., 2004. The Joint Evaluation of emergency assistance to Rwanda. *Humanitarian Exchange Magazine* Issue 26, March 2004. Available from: http://www.odihpn.org/humanitarian-exchange-magazine/issue-26/the-joint-evaluation-of-emergency-assistance-to-rwanda [Accessed 15 August 2012].
Bowerman, N. H. A., Frame, D. J., Huntingford, C., Lowe, J. A., and Allen, M. R., 2011. Cumulative carbon emissions, emissions floors policy and short-term rates of warming: implications for policy. *Philosophical Transactions of the Royal Society A*, 369, 45–66.
Bowonder, B., Kasperson, J. X., and Kasperson, R. E., 1993. Industrial risk management in India after Bhopal. *In*: S. Jasanoff, ed. *Learning from disaster: risk management after Bhopal*. Philadeplphia, PA: University of Pennsylvania Press, 66–90.
Bradfield, R., Wright, G., Burt, G., Cairns, G., and Heijdun, V. D., 2005. The origins and evolution of scenario techniques in long range business planning. *Futures*, 37, 795–812.
Brien, F., 2005. Building social capital. *Government Executive*, 37 (18), 82.
Brown, V. A. Harris, J. A., and Russell, J. Y., eds., 2010. *Tackling wicked problems through the transdisciplinary imagination*. London: Earthscan.
Brulle, J. R., Carmichael, J., and Jenkins, J. C., 2012. Shifting public opinion on climate change: an empirical assessment of factors influencing concern over climate change in the US, 2002–2010. *Climatic Change* [online], 114 (2), 169–188. Available from: http://www.springerlink.com/content/k17856khp026w174/fulltext.pdf [Accessed 9 October 2012].
Brundtland Commission, 1987. *Our common future*. Oxford, UK: Oxford University Press.
Buchanan-Smith, M., 2003. *How the Sphere Project came into being: a case study of policy-making in the humanitarian aid sector and the relative influence of research*. Working Paper 215. London: Overseas Development Institute.
Bulkeley, H. and Newell, P., 2010. *Governing climate change*. Oxford, UK: Routledge.
Bulkeley, H., Castán Broto, V., Hodson, M., and Marvin, S., eds., 2011. *Cities and low carbon transitions*. London: Routledge.
Burby, R., Deyle, R. E., Godschalk, D. R., and Olshansky, R. B., 2000. Creating hazard resilient communities through land-use planning. *Natural Hazards Review*, 1 (2), 99–106.
Burton, I., Kates, R. W., and White, G. F., 1993. *Environment as hazard*. London, UK: Guildford Press.
Busan Partnership for Effective Development Cooperation, 2011. Fourth High Level Forum on Aid Effectiveness, Busan, Republic of Korea, 29 November–1 December 2011. Available from: http://www.fao.org/fileadmin/user_upload/capacity_building/Busan_Effective_Development_EN.pdf [Accessed 13 August 2012].
Camilleri, J. A. and Falk, J., 2009. *Worlds in transition: evolving governance across a stressed plan*. Cheltenham, UK: Edward Elgar.
Cannon, T., Twigg, J., and Rowell, J., 2003. *Social vulnerability, sustainable livelihoods and disasters*. London: Benfield Hazard Research Centre. Available from: http://nirapad.org/admin/soft_archive/1308222298_Social%20Vulnerability-%20Sustainable%20Livelihoods%20and%20Disasters.pdf [Accessed 10 March 2013].
Carrington, D., 2010a. WikiLeaks cables reveal how US manipulated climate accord. *The Guardian*, 3 December 2010. Available from: http://www.guardian.co.uk/environment/2010/dec/03/wikileaks-us-manipulated-climate-accord [Accessed 10 December 2010].
Carrington, D., 2010b. Cancún deal leaves hard climate tasks to Durban summit in 2011, *The Guardian*, 14 December 2010.

Carson, R., 1962. *Silent spring*. Cambridge, MA: Houghton Mifflin.
Carter, W. N., 1991. *Disaster management: a disaster manager's handbook*. Manila: Asian Development Bank.
Cassidy, T., 2002. Problem-solving style, achievement motivation, psychological distress and response to a simulated emergency. *Counselling Psychology Quarterly*, 15 (4), 325–332.
Charlton, F. C., 1906. Humanitarianism, past and present. *International Journal of Ethics*, 17 (1), 48–55.
Charuvastra, A. and Cloitre, M., 2008. Social bonds and post-traumatic stress disorder. *Annual Review of Psychology*, 59, 301–328.
Crist, E., 2007. Beyond the climate crisis: a critique of climate change discourse. *Telos*, 141, 29–55.
Cutter, S. L., Barnes, L., Berry, M., Burton, C., Evans, E., Tate, E., and Webb, J., 2008. A place-based model for understanding community resilience to natural disasters. *Global Environmental Change*, 18, 598–606.
Daly, H., 1990. Sustainable growth: a bad oxymoron. *Environmental Carcinogenesis Reviews*, 8(2) 401–407.
Davies, K., 2012. Continuity, change and contest. meanings of 'humanitarian' from the 'religion of humanity' to the Kosovo war. HPG Working Paper. London: Overseas Development Institute.
DeFilippis, J., 2001. The myth of social capital in community development housing. *Policy Debate*, 12 (4), 781–806.
DeFilippis, J., 2002. Symposium on social capital: an introduction. *Antipode*, 34 (4), 790–795.
Department for International Development (DFID), 2000. *Sustainable livelihoods framework*. http://www.efls.ca/webresources/DFID_Sustainable_livelihoods_guidance_sheet.pdf [Accessed 22 August 2012].
Department for International Development (DFID), 2011a. *Humanitarian emergency response review*. Available from: http://www.dfid.gov.uk/Documents/publications1/HERR.pdf [Accessed 12 September 2012].
Department for International Development (DFID), 2011b. *Defining disaster resilience. A DFID Approach Paper*. Available from: http://www.dfid.gov.uk/Documents/publications1/Defining-Disaster-Resilience-DFID-Approach-Paper.pdf [Accessed 10 September 2012].
Dessai, S. and Schipper, E. L., 2003. The Marrakesh Accords and the Kyoto Protocol: analysis and future prospects. *Global Environmental Change*, 13, 149–153.
Dessai, S., Hulme, M., Lempert, R., and Pielke, R., 2009. Climate prediction: a limit to adaptation. *In*: N. Adger, I. Lorenzoni and K. O'Brien, eds. *Adapting to climate change: thresholds, values, governance*. New York, NY: Cambridge University Press, 64–78.
Dessler, A. E., 2012. *Introduction to modern climate change*. New York, NY: Cambridge University Press.
van Deth, J. W., 2003. Measuring social capital: orthodoxies and continuing controversies. *International Journal of Social Research Methodology*, 6 (1), 79–92.
Diamond, J. M., 2005. *Collapse: how societies choose to fail or succeed*. New York, NY: Viking Press.
Downing, T. E. and Lüdeke, M. K. B., 2002. International desertification: social geographies of vulnerability and adaptation. *In*: J. E. Reynolds and D. M. Stafford Smith, eds. *Global desertification*. Berlin: Dahlem University Press, 233–252.
Doyle, M. W., Stanley, E. H., Havlick, D. G., Kaiser, M. J., Steinbach, G., Graf, W. L., Galloway, G. E., and Riggsbee J. A., 2008. Aging infrastructure and ecosystem restoration. *Science*, 5861 (319), 286–287.

Dunn, S., 2002. *Reading the weathervane: climate policy from Rio to Johannesburg.* Worldwatch Institute, Worldwatch Paper 160. Available from: http://www.worldwatch.org/system/files/EWP160.pdf [Accessed 30 August 2012].

Durlauf, S. N., 1999. The case 'against' social capital. *Focus*, 20 (3), 1–5.

Dynes, R. and Rodriguez, H., 2006. *Finding and framing Katrina: the social construction of disaster.* Available from: http://understandingkatrina.ssrc.org/Dynes_Rodriguez/ [Accessed 3 September 2012].

Easterby-Smith, M. Burgoyne, J., and Araujo, L. eds., 1999. *Organizational learning and the learning organization.* London: Sage.

EIA, 2011. International Energy Outlook 2011, US Energy Information Administration. Available from: http://www.eia.gov/forecasts/ieo/pdf/0484 (2011).pdf [Accessed 5 September 2012].

Elahi, S., 2011. Here be dragons … exploring the 'unknown unknowns'. *Futures*, 43, 196–201.

Eldredge, N. and Gould S. J., 1972. Punctuated equilibria: an alternative to phyletic gradualism. *In*: T. J. M. Schopf, ed. *Models in paleobiology.* San Francisco, CA: Freemen, Cooper and Company, 82–115.

Ellerman, A. D. and Buchner, B., 2008. Over-allocation or abatement? A preliminary analysis of the EU ETS based on the 2005–06 emissions data. *Environmental and Resource Economics*, 41 (2), 267–287.

Ellerman, A. D., Convery, F. J., and de Perthuis, C., 2010. *Pricing carbon: the European Emissions Trading Scheme.* Cambridge, UK: Cambridge University Press.

Ellis, J., Winkler, H., Corfee-Morlot, J., and Gagnon-Lebrun, F., 2007. CDM: taking stock and looking forward. *Energy Policy*, 35, 15–28.

Elliston, J., 2004. Disaster in the making, *The Independent Weekly*, 22 September. http://www.indyweek.com/gyrobase/Content?oid=oid%3A22664 [Accessed 1 October 2012].

EU Commission, 2007. Communication from the Commission to the European Council and the European Parliament: an energy policy for Europe COM/2007/0001 final {SEC 2007. 12}. Available from: http://eur-lex.europa.eu/LexUriServ/site/en/com/2007/com2007_0001en01.pdf [Accessed 16 September 2012].

Eurobarometer, 2011. *Climate change: report.* Special Eurobarometer 372. Available from: http://ec.europa.eu/public_opinion/archives/ebs/ebs_372_en.pdf [Accessed 5 November 2012].

Evans, M. and Syrett, S., 2007. Generating social capital? The social economy and local economic development. *European Urban and Regional Studies*, 14 (1), 55–74.

Farr, J., 2004. Social capital: a conceptual history. *Political Theory*, 32 (1), 6–33.

Farrell, K. N., 2011. Snow White and the wicked problems of the west: a look at the lines between empirical description and normative prescription. *Science, Technology and Human Values*, 36 (3), 334–361.

Field J., 2008. *Social capital.* 2nd ed. London: Routledge.

Fine, B., 2001. *Social capital versus social theory: political economy and social science at the turn of the millennium.* Abingdon, UK: Routledge.

Fine, B., 2002. They f**k you up those social capitalists. *Antipode*, 34 (4), 796–799.

Fine, B., 2010. *Theories of social capital: researchers behaving badly.* London: Pluto Press.

Flood, R. L. and Romm, N. R. A., 1996. *Diversity management: triple loop learning.* San Francisco, CA: Wiley.

Folke, C., Hahn, T., Olsson, P., and Norberg, J., 2005. Adaptive governance as social–ecological systems. *Annual Review of Environmental Resources*, 30, 441–473.

Francis, J. A. and Vavrus, S. J., 2012. Evidence linking Arctic amplification to extreme weather in mid-latitudes. *Geophysical Research Letters*, 39, L06801. doi:10.1029/2012GL 051000. Available from: http://www.agu.org/journals/gl/gl1206/2012GL051000/2012 GL051000.pdf [Accessed 4 December 2012].

Friere, P., 2000. *Pedagogy of the oppressed*. New York, NY: Continuum.

Funtowicz, S. O. and Ravetz, J. R., 1991. A new scientific methodology for global environmental issues. *In*: R. Costanza, ed. *Ecological economics: the science and management of sustainability*. New York, NY: Columbia University Press, 137–152.

Fussel, H. M., 2005. *Vulnerability in climate change research: a comprehensive conceptual framework*. Breslauer Symposium. Paper 6. University of California International and Area Studies. Available from: http://repositories.cdlib.org/ucias/breslauer/6/ [Accessed 12 November 2012].

Gallopin, G. C., 1991. Human dimensions of global change: linking the global and the local processes. *International Social Science Journal*, 130, 707–718.

Gallopin, G. C., 2006. Linkages between vulnerability, resilience, and adaptive capacity. *Global Environmental Change*, 16, 293–303.

Galloway, G. E., Boesch, D. F., and Twilley, R. R., 2009. Restoring and protecting coastal Louisiana. *Issues in Science and Technology*, 2 (25) 29–38.

Galloway Jr, G. E., 2007. New directions in floodplain management. *Journal of the American Water Resources Association*, 3 (31), 351–357.

Giampietro, M., Mayumi, K., and Bukkens, S. G. F., 2001. Multiple-scale integrated assessment of societal metabolism: an analytical tool to study development and sustainability. *Environment, Development and Sustainability*, 3 (4), 275–307.

Giddens, A., 2009. *The politics of climate change*. Cambridge, UK: Polity Press.

Giddings, B., Hopwood, W., and O'Brien, G., 2002. Environment, economy and society: fitting them together into sustainable development. *Sustainable Development*, 10, 187–196.

Gilbertson, T. and Reyes, O., 2009. Carbon trading. How it works and why it fails. Occasional Paper Series Critical Currents. November 2009, No. 7. Uppsala: Dag Hammarskjöld Foundation.

Global Humanitarian Assistance (GHA), 2012a. GHA Report 2012: Summary, Development Initiative. Available from: http://www. globalhumanitarianassistance.org/wp-content/uploads/2012/07/GHA_Report_2012-Websingle.pdf [Accessed 14 March 2013].

Global Humanitarian Assistance (GHA), 2012b. GHA Report 2012. Available from: http:// www. globalhumanitarianassistance.org/wp-content/uploads/2012/07/GHAExecutive Summary2012Websingle.pdf [Accessed 1 July 2012].

Global Humanitarian Forum, 2009. Human impact report: climate change: the anatomy of a silent crisis, Geneva, 2009. Available from: http://www.bb.undp.org/uploads/file/ pdfs/energy_environment/CC%20human%20impact%20report.pdf [Accessed 10 July 2012].

Global Humanitarian Forum, 2010. The anatomy of a silent crisis: climate change human impact report, Global Humanitarian Forum, Geneva. Available from: http://www.eird. org/publicaciones/humanimpactreport.pdf [Accessed 28 August 2012].

Global Insight, 2007. A study of the European cosmetics industry. Executive summary. Report Prepared for the European Commission, Directorate General for Enterprise and Industry. Available from: http://ec.europa.eu/enterprise/newsroom/ cf/_getdocument.cfm?doc_id=4561 [Accessed 24 November 2012].

Goebbert, K., Jenkins-Smith, H. C., Klockow, K., Nowlin, M. C., and Silva, C. L., 2012. Weather, climate, and worldviews: the sources and consequences of public perceptions of changes in local weather patterns. *Weather, Climate and Society* [online], 4, 132–144. Available from: http://journals.ametsoc.org/doi/abs/10.1175/WCAS-D-11-00044.1 [Accessed 24 October 2012].

Goldenberg, S., 2009. China and US held secret talks on climate change deal, *The Guardian*, 18 May 2009. Available from: http://www.guardian.co.uk/world/2009/may/18/secret-us-china-emissions-talks [Accessed 23 October 2012].

Goldenberg, S., 2012. The inside story on climate scientists under siege, *The Guardian*, 17 February 2012. Available from: http://www.guardian.co.uk/environment/2012/feb/17/michael-mann-climate-war [Accessed 12 March 2012].

Gould, S. J. and Eldredge, N., 1977. Punctuated equilibria: the tempo and mode of evolution reconsidered. *Paleobiology*, 3 (2), 115–151.

Grasso, M., 2010. *Justice in funding adaptation under the International Climate Change regime.* New York, NY: Springer.

Great Britain Cabinet Office Civil Contingencies Secretariat, 2004. *Dealing with disasters*. Revised 3rd ed. London: The Stationery Office.

Green Climate Fund. http://gcfund.net/home.html

Green, H. and Fletcher, L., 2003. *The development of harmonised questions on social capital*. Office for National Statistics.

Grootaert, C. and van Bastelaer, T., 2001. *Understanding and measuring social capital: a synthesis of findings and recommendations from the Social Capital Initiative World Bank Social Capital Initiative*. Working Paper No 24. Washington, DC: The World Bank.

Grossman, G. M. and Krueger, A. B., 1995. Economic growth and the environment. *Quarterly Journal of Economics*, 110 (2), 353–377.

Grubb, M., 2011. Editorial: Durban: the darkest hour?. *Climate Policy*, 11, 1269–1271.

Gunderson, L. H. and Holling, C. S. (eds.), 2001. *Panarchy: understanding transformations in systems of humans and nature*. Washington, DC: Island Press.

Gupta, S., Tirpak, D. A., Burger, N., Gupta, J., Höhne, N., Boncheva, A. I., Kanoan, G. M., Kolstad, C., Kruger, J. A., Michaelowa, A., Murase, S., Pershing, J., Saijo, T., and Sari, A., 2007. Policies, instruments and co-operative arrangements. *In*: B. Metz, O. R. Davidson, P. R. Bosch, R. Dave, and L. A. Meyer, eds. *Climate Change 2007: Mitigation.* Contribution of Working Group III to the Fourth Assessment Report of the Intergovernmental Panel on Climate Change. Cambridge, UK: Cambridge University Press. Available from: http://www.ipcc.ch/pdf/assessment-report/ar4/wg3/ar4-wg3-chapter13.pdf [Accessed 18 September 2012].

Guston, D., 2001. Boundary organizations in environmental policy and science: an introduction. *Science, Technology, and Human Values*, 26 (4), 87–112.

Halpern, D., 2002. Why does social capital matter? Presentation for the Strategic Futures Group. Available from: http://www.cabinetoffice.gov.uk/media/cabinetoffice/strategy/assets/schalpern.pdf [Accessed 11 November 2012].

Halpern, D., 2005. *Social capital*. Cambridge, UK: Polity Press.

Hamilton, C., 2010. *Requiem for a species: why we resist the truth about climate change*. London: Earthscan.

Hansen, J., 2012. Game over for the climate, *New York Times*, 9 April 2012. Available from: http://www.nytimes.com/2012/05/10/opinion/game-over-for-the-climate.html?r=1 [Accessed 9 May 2012].

Hansen, J., Sato, M., and Ruedy, R., 2012. Perception of climate change. *Proceedings of the National Academy of Sciences* [online], 109 (37), 14726–14727. doi:10.1073/pnas.1205276109. Available from: http://pubs.giss.nasa.gov/docs/2012/2012_Hansen_etal_1.pdf [Accessed 28 August 2012].

Hansen, J., Sato, M., and Ruedy, R., 2012. Perception of climate change. *Proceedings of the National Academy of Sciences*, 109, 14726–14727.

Harpham, T., Grant, E., and Thomas, E., 2002. Measuring social capital within health surveys: key issues. *Health Policy and Planning*, 17 (1), 106–111.

Harrabin, R., 2010. UN Climate talks in China end without breakthrough, *BBC News*, 9 October 2010. Available from: http://www.bbc.co.uk/news/science-environment-11508913 [Accessed 1 August 2012].

Harrison, F., 1879. Science and humanity. *The North American Review*, 129, 322–342.

Harvey, D., 2008. On the deep relevance of a certain footnote in Marx's capital. *Human Geography*, 1 (2), 26–31.

Harvey, P., 2009. *Towards good humanitarian governance: the role of the affected state in disaster response.* HPG Policy Brief 37.

Hassing, P., O'Brien, G., O'Keefe, P., and Tellam I., 2009. Climate change adaptation: what's the bottom line. *COP15 Press Release*, 16 December 2009, Copenhagen.

Hatzius, T. 1996. *Sustainability and institutions – catchwords or new agenda for ecologically sound development?* Institute of Development Studies Working Paper 48.

Her Majesty's Government, 2010. Government Response to the House of Commons Science and Technology Committee 8th Report of Session 2009–10. The disclosure of climate data from the Climatic Research Unit at the University of East Anglia. Available from: http://www.official-documents.gov.uk/document/cm79/7934/7934.pdf [Accessed 1 December 2012].

Hewitt, K., 1983. *Interpretations of calamity: from the viewpoint of human ecology.* London: Unwin Hyman.

Hewitt, K., 1998. Excluded perspectives in the social construction of disaster. *In*: E. Quarantelli, ed. *What is a disaster? Perspectives on the question.* New York, NY: Routledge, 75–91.

Hickman, L., 2012. Climate change study forces sceptical scientists to change minds, *The Guardian*, 29 July 2012. Available from: http://www.guardian.co.uk/science/2012/jul/29/climate-change-sceptics-change-mind [Accessed 1 August 2012].

Hilhorst, D. and Bankoff, G., 2004. Introduction: mapping vulnerability. *In*: G. Bankoff, G. Frerks, and D. Hilhorst, eds. 2004. *Mapping vulnerability: disasters, development and people.* London: Earthscan, 1–24.

Holgate, M. W., Kassas, M., and White, G. F., 1982. *The world environment, 1972–1982.* Dublin: Tycooly International Publishing, Ltd.

Holland, J., 1995. *Hidden order: how adaptation builds complexity.* Reading, MA: Perseus Books.

Holling, C. S., 1973. Resilience and stability of ecological systems. *Annual Review of Ecology and Systematics* [online], 4, 1–23. Available from: http://www.jstor.org/view/00664162/di975347/97p0062a/0 [Accessed 9 August 2012].

Holling, C. S., 2001. Understanding the complexity of economic, ecological, and social systems. *Ecosystems,* 2001 (4), 390–405.

Holling, C. S., 2004. From complex regions to complex worlds. *Ecology and Society* [online], 9 (1), 11. Available from: http://www.ecologyandsociety.org/vol9/iss1/art11 [Accessed 18 September 2012].

Hopwood, B., Mellor, M., and O'Brien, G., 2005. Sustainable development: mapping different approaches. *Sustainable Development*, 13, 38–52.

House of Lords, 2008. House of Lords European Union Committee 33rd Report of Session 2007–08. *The Revision of the EU's Emissions Trading System: Report with Evidence.* London: The Stationary Office. Available from: http://www.publications.parliament.uk/pa/ld200708/ldselect/ldeucom/197/197.pdf [Accessed 5 November 2012].

Huq, S., 2010. Bonn climate talks: picking up the pieces after Copenhagen, *The Guardian*, 12 April 2010. Available from: http://www.guardian.co.uk/environment/cif-green/2010/apr/12/bonn-climate-talks-copenhagen?intcmp=239 [Accessed 12 April 2010].

Huss, W. R., 1988. A move toward scenario analysis. *International Journal of Forecasting*, 4, 377–388.

Hutchinson, J., 2004. Social capital and community building in the inner city. *Journal of the American Planning Association*, 70 (2), 168–175.

Hyogo Declaration, 2005. World Conference on Disaster Reduction 18–22 January 2005, Kobe, Hyogo, Japan. Available from: http://www.unisdr.org/wcdr/intergover/official-doc/L-docs/Hyogo-declaration-english.pdf [Accessed 21 August 2012].

Hyogo Framework for Action, 2005. Hyogo Framework for Action 2005–2015: building the resilience of nations and communities to disasters. World Conference on Disaster Reduction 18–22 January 2005, Kobe, Hyogo, Japan. Available from: http://www.unisdr.org/wcdr/intergover/official-doc/L-docs/Hyogo-framework-for-action-english.pdf [Accessed 22 August 2012].

IISD RS, 2010. Summary of the Bonn Climate Change Talks: 31 May–11 June 2010. Available from: http://www.iisd.ca/vol12/enb12472e.html [Accessed 24 August 2012].

Inderberg, T. H. and Eikeland, P. O., 2009. Limits to adaptation: analysing institutional constraints. *In*: N. Adger, I. Lorenzoni, and K. O'Brien, eds. *Adapting to climate change: thresholds, values, governance*. Cambridge, UK: Cambridge University Press.

Intergovernmental Panel on Climate Change (IPCC), 1990. *Policymakers summary of the formulation of response strategies*. Report prepared for the IPCC by Working Group III, IPCC.

Intergovernmental Panel on Climate Change (IPCC), 2001. *Third Assessment Report (TAR): impacts, adaptation and vulnerability*. Available from: http://www.grida.no/climate/ipcc_tar/ [Accessed 13 December 2012].

Intergovernmental Panel on Climate Change (IPCC), 2004. *16 years of scientific assessment in support of the Climate Convention*. Geneva: IPCC.

Intergovernmental Panel on Climate Change (IPCC), 2007a. Climate change 2007: synthesis report. *In*: R. K. Pachauri, and A. Reisinger, eds. *Contribution of Working Groups I, II and III to the Fourth Assessment Report of the Intergovernmental Panel on Climate Change*. Geneva: IPCC.

Intergovernmental Panel on Climate Change (IPCC), 2007b. *Climate change 2007: synthesis report*. Cambridge, UK: Cambridge University Press.

Intergovernmental Panel on Climate Change (IPCC), 2012a. Summary for policymakers. *In*: C. B. Field, V. Barros, T. F. Stocker, D. Qin, D. J. Dokken, K. L. Ebi, M. D. Mastrandrea, K. J. Mach, G.-K. Plattner, S. K. Allen, M. Tignor, and P. M. Midgley, eds. *Managing the risks of extreme events and disasters to advance climate change adaptation. A Special Report of Working Groups I and II of the Intergovernmental Panel on Climate Change*. Cambridge, UK: Cambridge University Press, 1–19.

Intergovernmental Panel on Climate Change (IPCC), 2012b. *Managing the risks of extreme events and disasters to advance climate change adaptation. A Special Report of Working Groups I and II of the Intergovernmental Panel on Climate Change*, C. B. Field, V. Barros, T. F. Stocker, D. Qin, D. J. Dokken, K. L. Ebi, M. D. Mastrandrea, K. J. Mach, G.-K. Plattner, S. K. Allen, M. Tignor and P. M. Midgley, eds. Cambridge, UK: Cambridge University Press, 582 pp.

International Council for Local Environmental Initiatives (ICLEI), 2010. Local government climate roadmap. Available from: http://www.iclei.org/index.php?id=11383 [Accessed 20 August 2012].

International Energy Agency (IEA), 2002. World Energy Outlook. Paris: OECD/IEA. Available from: http://www.worldenergyoutlook.org/media/weowebsite/2008-1994/weo2002_part1.pdf [Accessed 14 March 2013].

International Energy Agency (IEA), 2003. Energy to 2050: Scenarios for Sustainable Futures IEA/OECD, Paris. Available from: http://isulibrary.isunet.edu/opac/doc_num.php?explnum_id=348 [Accessed 14 March 2013].

International Energy Agency (IEA), 2009. IEA offers blueprint to deliver on ambitious climate change goals and urges all governments to send a strong signal to spur new investment for clean energy. IEA Press Release 14 December 2009. Available from: http://www.iea.org/newsroomandevents/pressreleases/2009/december/name,20278,en.html [Accessed 14 March 2013].

International Energy Agency (IEA), 2011a. World Energy Outlook, Executive Summary. Available from: http://www.iea.org/Textbase/npsum/weo2011sum.pdf [Accessed 5 January 2013].

International Energy Agency (IEA), 2011b. CO_2 emissions from fuel combustion: highlights. Available from: http://www.iea.org/co2highlights/co2highlights.pdf [Accessed 29 September 2012].

Jacobs, J., 1961. *The death and life of great American cities.* New York, NY: Random House.

Janssen, M. A., Schoon, M. L., Ke, W., and Borner, K., 2006. Scholarly networks on resilience, vulnerability and adaptation within the human dimensions of global change. *Global Environmental Change*, 16, 240–252.

Johnston, D. and Paton, D., 2001. Disasters and communities: vulnerability, resilience and preparedness. *Disaster and Prevention Management*, 10 (4), 270–277.

Johnston, G. and Percy-Smith, J., 2003. In search of social capital. *Policy & Politics*, 31 (3), 321–334.

Johnston, R. J. and Sidaway, J. D., 2004. *Geography & geographers: Anglo–American human geography since 1945.* London: Arnold.

Kahan, D., 2012. Why we are poles apart on climate change. *Nature*, 488, 255. Available from: http://www.nature.com/polopoly_fs/1.11166!/menu/main/topColumns/topLeftColumn/pdf/488255a.pdf [Accessed 3 September 2012].

Karoly, D. J., Braganza, K., Stott, P. A., Arblaster, J. M., Meehl, G. A., Broccoli, A. J., and Dixon, K. W., 2003. Detection of a human influence on North American climate. *Science*, 302, 1200–1203.

Kastenhofer, K., 2011. Risk assessment of emerging technologies and post-normal science. *Science, Technology and Human Values*, 36 (3), 307–333.

Kates, R. W., 1962. *Hazard and choice perception in floodplain management.* Paper 78, Department of Geography Research. Chicago, IL: University of Chicago Press.

Kates, R. W., 1971. Natural hazard in human ecological perspective: hypotheses and models. *Economic Geography*, 47 (3), 438–451.

Kates, R. W., Colten, C. E., Laska, S., and Leatherman, S. P., 2006. Reconstruction of New Orleans after hurricane Katrina: A research perspective. *Proceedings of the National Academy of Sciences*, (103), 14653–14660.

Kaufman, S., 2006. The criminalization of New Orleanians in Katrina's wake: understanding Katrina: perspectives from the social sciences. Available from: http://understandingkatrina.ssrc.org/Kaufman/

Keen, M., Brown, V. A., and Dyball, R., 2005. Social learning: a new approach to environmental management. *In*: M. Keen, V. A. Brown, R. Dyball, eds. *Social learning in environmental management: towards a sustainable future.* London: Earthscan.

Kelly, R., Sirr, L., and Ratcliffe, J., 2004. Futures thinking to achieve sustainable development at local level in Ireland. *Foresight*, 6 (2), 80–90.

Kett, M., 2010. Humanitarian disaster relief: disability and the new sphere guidelines. Available from: http://www.ucl.ac.uk/global-disability-research/downloads/Maria_Kett_Sphere_Futures_Presentation.pdf [Accessed 16 August 2012].

Kirkby, S. J. and Moyo, S., 2001. Environmental security, livelihoods and entitlement. *In*: N. Middleton, P. O'Keefe, R. Visser, eds. *Negotiating poverty new directions. Renewed Debate.* London: Pluto Press, 148–161.

Kriegman, O., Great Transition Initiative (lead authors), 2007. Global citizens movement. *In*: C. J. Cleveland, ed. *Encyclopedia of Earth.* Washington, DC: Environmental Information Coalition, National Council for Science and the Environment. [first published in *Encyclopedia of Earth* 9 November 2007; last revised date 9 November 2007]. Available from: http://www.eoearth.org/article/Global_citizens_movement [Accessed 11 September 2012].

Krishna, A. and Shrader, E., 1999. *Social capital assessment tool conference on social capital and poverty reduction.* Prepared for Conference on Social Capital and Poverty Reduction, The World Bank, Washington, DC, 22–24 June 1999. Available from: http://siteresources.worldbank.org/INTSOCIALCAPITAL/Resources/Social-Capital-Assessment-Tool–SOCAT-/sciwp22.pdf [Accessed 10 September 2012].

Kuhn, T., 1962. *The structure of scientific revolutions.* Chicago, IL: University of Chicago Press.

Lagadec, P., 2005. *Crisis Management in the 21st Century: 'Unthinkable' Events in 'Inconceivable' Contexts.* Paper presented at Centre National de la Recherche Scientifique, France: 2005. Available from: http://ceco.polytechnique.fr/fichiers/ceco/publications/pdf/2005-03-14-219.pdf [Accessed 13 September 2012].

Lewandowsky, S., 2011. Why Australia is vulnerable to both climate change and climate sceptics, *The Guardian*, 15 June 2011. Available from: http://www.guardian.co.uk/environment/blog/2011/jun/15/climate-change-skeptic-australia [Accessed 15 June 2011].

Li, J., Pickles, A., and Savage, M., 2003. *Social capital dimensions, social trust and quality of life in Britain in the late 1990s.* ISER Seminar Paper, University of Essex, Colchester, UK.

Liebmann, M. and Pavanello, S., 2007. *A critical review of the knowledge and education indicators of community-level disaster risk reduction for the Benfield UCL Hazard Research Centre (BUHRC).* London: BUHRC.

Ljungman, C. M., 2004. *Applying a rights-based approach to development: concepts and principles.* Conference paper presented at The Winners and Losers from Rights Based Approaches to Development, 21–22 February 2005, Manchester, UK.

Long Martello, M. and Iles, A., 2006. Making climate change impacts meaningful: framing, methods, and process in coastal zone and agricultural assessments. *In*: A. E. Farrell and J. Jäger, eds. *Assessments of regional and global environmental risks. Designing processes for the effective use of science in decision making.* Washington, DC: Resources for the Future, 101–118.

Loury, G., 1977. A dynamic theory of racial income differences in Wallace. *In*: P. A. Wallace and A. M. LaMond, eds. *Women, minorities and employment discrimination.* Lexington, MA: Lexington Books.

Lovelock, J. E., 1979. *Gaia: a new look at life on Earth.* Oxford, UK: Oxford University Press.

Lovelock, J. E., 2006. *The revenge of Gaia.* London: Penguin.

Lovins, A. B. and Lovins, L. H., 1982. *Brittle power: energy strategy for national security.* Baltimore, MD: Brick House Publishing. Available from: http://library.uniteddiversity.

coop/Energy/BrittlePower.pdf [Accessed 10 March 2013].
Lynch, J., Due, P., Muntaner, C., and Davey Smith, G., 2000. Social capital – is it a good investment strategy for public health? *Journal of Epidemiology and Community Health*, 54, 404–408.
Mace, M. J., 2005. Funding for Adaptation to Climate Change: UNFCCC and GEF Developments since COP-7. *Review of European Community and International Environmental Law*, 14 (3), 225–246.
Manyena, S. B., 2006. The concept of resilience revisited. *Disasters*, 30 (4), 433–450.
de Marchi, B. and Ravetz, J. R., 1999. Risk management and governance: a post-normal science approach. *Futures*, 31 (7), 743–757.
McCarthy, J. J., Canziani, O. F., Leary, N. A., Dokken, D. J., and White, K. S. eds., 2001. *Climate change 2001: impacts, adaptation, vulnerability*. Contribution of Working Group 2 to the Third Assessment Report of the Intergovernmental Panel on Climate Change. Cambridge, UK: Cambridge University Press.
McElroy, M. W., 2000. Integrating complexity theory, knowledge management and organizational learning. *Journal of Knowledge Management* [online], 4 (3), 195–203. MCB University Press. Available from: http://www.macroinnovation.com/images/ IntegratingandOL.pdf [Accessed 1 October 2012].
McEntire, D. A., 2001. Triggering agents, vulnerabilities and disaster reduction: towards a holistic paradigm. *Disaster Prevention and Management*, 10, 189–196.
McEntire, D. A. and Fuller, C., 2002. The need for a holistic theoretical approach: an examination from the El Niño disasters in Peru. *Disaster Prevention and Management*, 11 (2), 128–140.
McEntire, D. A., Fuller, C., Johnston, C. W., and Weber, R., 2003. A comparison of disaster paradigms: the search for a holistic policy guide. *Public Administration Review*, 62 (3), 267–281.
McGuire, B., 2012. *Waking the giant: how a changing climate triggers earthquakes, tsunamis and volcanoes*. Oxford, UK: Oxford University Press.
McKibben, B., 2012. The great carbon bubble, *Aljazeera*, 14 February 2012. Available from: http://www.aljazeera.com/indepth/opinion/2012/02/20122117544271875.html [Accessed 1 October 2012].
McKie, R., 2012. Rate of Arctic summer sea ice loss is 50% higher than predicted, *The Guardian*, 11 August 2012. Available from: http://www.guardian.co.uk/environment/ 2012/aug/11/arctic-sea-ice-vanishing?intcmp=122 [Accessed 11 August 2012].
McManus, P., 2000. Environmental hazard. *In*: R. J. Johnston, D. Gregory, G., Pratt, and M. Watts, eds. *The dictionary of human geography*. 4th ed. London: Blackwell, 216–217.
Middleton, N. and O'Keefe, P., 2000. *Redefining sustainable development*. London: Pluto Press.
Middleton, N. and O'Keefe, P., 2003. *Rio plus ten: politics, poverty and environment*. London: Pluto Press.
Middleton, N., O'Keefe, P. and Moyo, S., 1994. *Tears of the crocodile: from reality in the developing world*. London: Pluto Press.
Mileti, D., 1999. *Disasters by design: a reassessment of natural hazards in the United States*. Washington, DC: Joseph Henry Press.
Miller, T., 2009. Experts sound off on US role, expectations for Copenhagen. *PBS Newshour*, 25 November 2009. Available from: http://www.pbs.org/newshour/updates/ north_america/july-dec09/climate_11-25.html [Accessed 3 October 2012].
Miller, M., de le Rosa, E., and Bohn, M., 2008. *The challenge of moving from acknowledgement to action: a review of vulnerability to environmental stresses and natural hazards in PRSPs*. Stockholm: Stockholm Environment Institute.

Mintzberg, H., Ahlstrand, B., and Lampel, J., 1998. *Strategy safari: a guided tour through the wilds of strategic management*. New York, NY: Prentice Hall.

Modi, V., McDade, S., Lallement, D., and J. Saghir., 2006. *Energy and the Millennium Development Goals*. New York, NY: Energy Sector Management Assistance Programme, United Nations Development Programme, UN Millennium Project, and World Bank. Available from: http://www.unmillenniumproject.org/documents/MP_Energy_Low_Res.pdf [Accessed 12 October 2012].

Mol, A. P. J., 2012. Carbon flows, financial markets and climate change mitigation. *Environmental Development*, 1, 10–24.

Monnier, J., Hobfoll, S. E., Dunnahoo, C. L., Hulsizer, M. R., and Johnson R., 1998. There is more than rugged individualism in coping. Part 2: construct validity and further model testing. *Anxiety, Stress and Coping*, 11, 247–272.

Moore, S., Daniel, M., Linnan, L., Campbell, M., Benedict, S., and Meier, A., 2004. After Hurricane Floyd passed: investigating the social determinants of disaster preparedness and recovery. *Family and Community Health*, 27 (3), 204–217.

Murphy, B. L., 2007. Locating social capital in resilient community-level emergency management. *Natural Hazards*, 41, 297–315.

Nakagawa, Y. and Shaw, R., 2004. Social capital: a missing link to disaster recovery. *International Journal of Mass Emergencies and Disasters*, 22 (1), 5–34.

Navarro, V., 2002. Politics, power and quality of life: a critique of social capital. *International Journal of Health Services*, 32 (3), 423–432.

Navarro, V., 2004. Commentary: Is social capital the solution or the problem? *International Journal of Epidemiology*, 33, 672–674.

Nilsson, A. E. and Swartling A. G., 2009. *Social learning about climate adaptation: global and local perspectives*. Working Paper. Stockholm: Stockholm Environment Institute, Available from: http://www.sei-international.org/mediamanager/documents/Publications/Policy-institutions/social_learning_wp_091112.pdf [Accessed 14 November 2012].

Norris, F. H., Stevens, S. P., Pfefferbaum, B., Wyche, K. F., and Pfefferbaum, R. L., 2008. Community resilience as a metaphor, theory, set of capacities, and strategy for disaster readiness. *American Journal of Community Psychology*, 41, 127–150.

O'Brien, G., 2006. The globalisation of disaster: UK emergency preparedness: a step in the right direction? *Journal of International Affairs*, 64 (5), 63–85.

O'Brien, G., 2009. Resilience and vulnerability in the European energy system. *Energy and Environment*, 20 (3), 399–410.

O'Brien, G. and Read, P., 2005. Future UK emergency management: new wine, old skin? *Disaster Prevention and Management*, 14 (3), 353–361.

O'Brien, G. and O'Keefe, P., 2006. The future of nuclear power in Europe: a response. *International Journal of Environmental Studies*, 63, 121–130.

O'Brien, G. and Hope, A., 2010. Localism and energy: negotiating approaches to embedding resilience in energy systems. *Energy Policy* [online], 38 (12), 7550–7558. Available from: http://dx.doi.org/10.1016/j.enpol.2010.03.033 [Accessed 9 November 2012].

O'Brien, G. and O'Keefe, P., 2010. Resilient responses to climate change and variability: a challenge for public policy. *The International Journal of Public Policy*, 6 (3–4), 369–385.

O'Brien, G., O'Keefe, P., Rose, J., and Wisner, B., 2006. Climate change and disaster management. *Disasters*, 30 (1), 64–80.

O'Brien, G., O'Keefe, P., and Rose, J., 2007. Energy, poverty and governance. *International Journal of Environmental Studies*, 64, (5), 607–618. DOI: 10.1080/00207230600841385.

O'Brien, G., O'Keefe, P., Meena, H., Rose, J., and Wilson, L., 2008. Climate adaptation from a poverty perspective. *Climate Policy*, 8 (2), 194–201.

O'Brien, G., O'Keefe, P., Gadema, Z., and Swords, J., 2010. Approaching disaster management through social learning. *Disaster Prevention and Management*, 19 (4), 498–508.

O'Brien, G., O'Keefe, P., and Devisscher, T. eds., 2011. *The adaptation continuum: groundwork for the future.* Saarbrucken, Germany: Lambert Academic Publishing.

O'Brien, G., Bhatt, M., Saunders, W., Gaillard, J. C., and Wisner, B., 2012. Local government and disaster. *In*: B. Wisner, J. C. Gaillard, I. Kelman, eds. *The Routledge handbook of hazards and disaster risk reduction.* Oxford, UK: Routledge, 629–640.

O'Keefe, P., Westgate, K., and Wisner, B., 1976. Taking the naturalness out of natural disasters. *Nature*, 260, 566–7.

O'Keefe, P., O'Brien, G., and Pearsall, N., 2010. *The future of energy use.* London: Earthscan.

Oliver, J. G. J., Janssens-Maenhout, G., and Peters, J. A. H. W., 2012. *Trends in global CO_2 emissions. 2012 Report.* The Hague: PBL Netherlands Environmental Assessment Agency. Available from: http://www.pbl.nl/sites/default/files/cms/publicaties/PBL_2012_Trends_in_global_CO2_emissions_500114022.pdf [Accessed 15 August 2012].

Opensecrets.org, 2012. Energy/natural resources: top contributors to federal candidates, parties, and outside groups. Available from: http://www.opensecrets.org/industries/contrib.php?cycle=2012&ind=E [Accessed 11 October 2012].

Opperman, J. J., Galloway, G., Fargione, E. J., Mount, J. F., Richter, B. D., and Secchi, S., 2009. Sustainable floodplains through large-scale reconnection to rivers. *Science*, 326, 1487–1488.

Ormrod, J. E., 1999. *Human learning.* 3rd ed. Upper Saddle River, NJ: Prentice Hall.

Ott, H. E., Bouns, B., Sterk, W., and Witneben, B., 2005. It takes two to tango – climate policy at COP10 in Buenos Aires and beyond. *Journal for European Environmental and Planning Law*, 2, 84–91.

Oxfam, 2007. *Adapting to climate change. What is needed in poor countries and who should pay?* Oxfam Briefing Paper 104.

Paldam, M., 2000. Social capital: one or many? Definition and measurement. *Journal of Economic Surveys*, 14 (5), 629–653.

Parmesan, C. and Yohe, G., 2003. A globally coherent fingerprint of climate change impacts across natural systems. *Nature*, 421 (6918) 37–42.

Parnell, S., Simon, D., and Vogel, C., 2007. Global environmental change: conceptualising the growing challenge for cities in poor countries. *Area*, 39 (3), 357–369.

Parry, M., Arnell, N., Berry, P., Dodman, D., Fankhauser, S., Hope, C., Kovats, S., Nicholls, R., Satterthwaite, D., Tiffin, R., and Wheeler, T., 2009. *Assessing the costs of adaptation to climate change: a review of the UNFCCC and other recent estimates.* London: International Institute for Environment and Development and Grantham Institute for Climate Change. Available from: http://pubs.iied.org/pdfs/11501IIED.pdf [Accessed 15 August 2012].

Paton, D. and Jackson, D., 2002. Developing disaster management capability: an assessment centre approach. *Disaster Prevention and Management*, 11 (2), 115–122.

Paton, D., Smith, L., and Violanti J., 2000. Disaster response: risk, vulnerability and resilience. *Disaster Prevention and Management*, 9 (3), 173–179.

Pelling, M., 1998. Participation, social capital and vulnerability to urban flooding in Guyana. *Journal of International Development*, 10, 469–486.

Pelling, M. and High, C., 2005. Understanding adaptation: what can social capital offer assessments of adaptive capacity? *Global Environmental Change*, 15 (4), 308–319.

Perrow, C., 1999. *Normal accidents: living with high risk technologies.* Princeton, NJ: Princeton University Press.

Perry, R. W. and Lindell, M. K., 2003. Preparedness for emergency response: guidelines for the emergency planning process. *Disasters*, 27 (4), 336–350.
Perry, R. W. and Peterson, D. M., 1999. The impacts of disaster exercises on participants. *Disaster Prevention and Management*, 8 (4), 241–254.
Petersen, A. C., Cath, A., Hage, M., Kunsler, E., and van der Sluijs, J. P., 2011. Post-normal science in practise at the Netherlands Environmental Assessment Agency. *Science, Technology and Human Values*, 36 (3), 362–388.
Pierson, D. and Tankersley, J., 2009. China's climate pledge raises expectations for Copenhagen summit. *Los Angeles Times*, 27 November 2009. Available from: http://articles.latimes.com/2009/nov/27/world/la-fg-china-climate28-2009nov28 [Accessed 16 October 2012].
Pitt, M., 2008. *The Pitt Review: learning lessons from the 2007 floods*. Cabinet Office [online] http://webarchive.nationalarchives.gov.uk/20100807034701/http://archive.cabinetoffice.gov.uk/pittreview/thepittreview.html [Accessed 3 December 2012].
Policy Research Initiative (PRI), 2005. *Social capital as a public policy tool. Project Report*. Canada: Policy Research Initiative (PRI). Available from: http://www.horizons.gc.ca/doclib/PR_SC_SocialPolicy_200509_e.pdf [Accessed 10 March 2013].
Portes, A., 2000. The two meanings of social capital. *Sociological Forum*, 15 (1), 1–12.
Portes, A. and Landolt, P., 1996. The downside of social capital. *The American Prospect*, 26, 18–21.
Putnam, R. D., 1993. The prosperous community: social capital and public life. *The American Prospect*, 13, 35–42.
Putnam, R. D., 1995. Bowling alone: America's declining social capital. *Journal of Democracy*, 6 (1), 65–78.
de Puydt, P. E., 1860. *Panarchy*. First published in French in Revue Trimestrielle, Brussels. English version http://www.panarchy.org/depuydt/1860.eng.html [Accessed 7 December 2012].
Quarantelli, E. L., 1992. The case for a generic rather than agent-specific approach to disasters. *Disaster Management*, 2, 191–196.
Quarantelli, E. L., 2005. *Disaster planning, emergency management and civil protection: the historical development of organized efforts to plan for and to respond to disasters*. University of Delaware Library. Available from: http://dspace.udel.edu:8080/dspace/handle/19716/673 [Accessed 20 October 2012].
Quarantelli, E. L., Russell, R., and Dynes, R. R., 1972. When disaster strikes (it isn't much like what you've heard and read about). *Psychology Today*, 5 (9), 66–70.
Raskin, P., Banuri, T., Gallopin, G., Gutman, P., Hammond, A., Kates, R., and Swart, R., 2002. Great transition: the promise and lure of the times ahead. Available from: http://tellus.org/documents/Great_Transition.pdf [Accessed 24 November 2012].
Redclift, M., 2005. Sustainable development (1987–2005): an oxymoron comes of age. *Sustainable Development* [online], 13, 212–227. Available from: http://www.homepages.ucl.ac.uk/~ucessjb/S3%20Reading/redclift%202005.pdf [Accessed 19 November 2012].
Reid, H. and Huq, S., 2007. How we are set to cope with the impacts. IIED Briefing. London: International Institute of Environment and Development.
REN21, 2010. Renewables 2010 Global Status Report (Paris: REN21 Secretariat). Available from: http://www.ren21.net/Portals/0/documents/activities/gsr/REN21_GSR_2010_full_revised%20Sept2010.pdf [Accessed 14 March 2013].
REN21, 2011. Global Status Report, REN21. Available from: http://www.globalpolicy.org/images/pdfs/1_-_Report_-_REN_21.pdf [Accessed 14 March 2013].

Resilience Alliance, 2011. Key concepts. http://www.resalliance.org/index.php/key_concepts [Accessed 14 March 2013].

Rhodes, F., 2011. Flood warnings for infrastructure: tailored flood warning services. *Fifth International Conference on Flood management (ICFM5)*, 27–29 September 2011, Tsukuba, Japan.

Ritchey, T., 2007. Wicked problems: structuring social messes with morphological analysis. Swedish Morphological Society. Available from: http://www.swemorph.com/pdf/wp.pdf [Accessed 17 August 2012].

Root, T. L., Price J. T., Hall, K. R., Schneider, S. H., Rosenzweig, C., and Pounds, J. A., 2003. Fingerprints of global warming on wild animals and plants. *Nature*, 421, 57–60.

Rostow, W. W., 1960. *The stages of economic growth: a non-communist manifesto*. Cambridge, UK: Cambridge University Press.

Rotter, J. B., 1982. *The development and application of social learning theory*. New York, NY: Praeger.

Safran, P., 2003. A strategic approach for disaster and emergency assistance. Contribution to the 5th Asian Disaster Reduction Centre International Meeting and the 2nd UNISDR Asian Meeting, 15–17 January 2003, Kobe, Japan.

Sardar, Z., 2009. Welcome to postnormal times. doi:10.1016/j.futures.2009.11.028 (article in press).

Savage, M., 2002. Business Continuity Planning. *Work Study*, 51 (5), 254–261.

Schaafstal, A. M., Johnston, J. H., and Oser, R. L., 2001. Training teams for emergency management. *Computers in Human Behavior*, 17, 615–26.

Schipper, E. L. F., 2006. Conceptual history of adaptation in the UNFCCC process. *Review of European Community and International Environmental Law*, 15 (1), 82–92.

Schipper, L. and Pelling, M., 2006. Disaster risk, climate change and international development: scope for, and challenges to, integration. *Disasters*, 30 (1), 19–38.

Schultz, J. M., Russell, J., and Espinel, Z., 2005. Epidemiology of tropical cyclones: the dynamics of disaster, disease and development. *Epidemiological Reviews* [online], 27, 21–35. Available from: http://epirev.oxfordjournals.org/cgi/reprint/27/1/21 [Accessed 24 October 2012].

Schwartz, P., 1991. *The art of the long view*. New York, NY: Currency Doubleday.

Seifert, J. W., 2002. The effects of September 11, 2001, terrorist attacks on public and private information infrastructures: a preliminary assessment of lessons learned. *Government Information Quarterly*, 19, 225–42.

Sen, A., 1982. *Poverty and famines: an essay on entitlements and deprivation*. Oxford, UK: Clarendon Press.

Sen, A., 1999. *Development as freedom*. Oxford, UK: Oxford University Press.

Senge, P., 1990. Building learning organizations. *In*: D. S. Pugh, ed. (1997) *Organization theory*. 4th ed. New York, NY: Penguin.

Sewell, J. P. and Salter, M. B., 1995. Panarchy and other norms for global governance: Boutros-Ghali, Rosenau, and beyond. *Global Governance*, 1 (3), 373–82.

Sheppard, K., 2009. Obama's Copenhagen stopover: The US president's cameo appearance at Copenhagen's climate summit might make more of an impact than his critics realise, *The Guardian*, 26 November 2009. Available from: http://www.guardian.co.uk/commentisfree/cifamerica/2009/nov/26/copenhagen-barack-obama [Accessed 12 September 2012].

Sheppard, S. R. J., Shaw, A., Flanders, D., Burch, S., Wiek, A., Carmichael, J., Robinson, J., and Cohen, S., 2011. Future visioning of local climate change: a framework for community engagement and planning with scenarios and visualization. *Futures*, 43, 400–412.

Shoothill, 2012. Facebook Flood Alerts. Available from: http://www.shoothill.com/floodalerts/ [Accessed 10 December 2012].
Simon, H. A., 1957. *Models of man – social and rational*. New York, NY: Wiley.
Smith, K. P. and Christakis N. A., 2008. Social networks and health. *Annual Review of Sociology*, 34, 405–429.
Smith, N., 1984. *Uneven development, nature, capital and the production of space*. Oxford, UK: Basil Blackwell.
Smith, N. and O'Keefe, P., 1980. Geography, Marx and the concept of nature. *Antipode*, 12, 30–39.
Smith, N. and O'Keefe, P., 1985. Geography, Marx and the concept of nature. *Antipode*, 17 (2–3), 79–88.
Sperling, F. and Szekely, F., 2005. Disaster risk management in a changing climate. Informal discussion paper prepared for the World Conference on Disaster Reduction on behalf of the Vulnerability and Adaptation Resource Group (VARG), Washington, DC.
Stern, N., 2006. *The Stern Review: the economics of climate change*. London: HM Treasury.
Stratton, A. and Vidal, J., 2009. Connie Hedegaard resigns as president of Copenhagen climate summit, *The Guardian*, 16 December 2009. Available from: http://www.guardian.co.uk/environment/2009/dec/16/connie-hedegaard-copenhagen-resigns [Accessed 2 August 2012].
Sustainable Development Commission (SDC), 2009. *Prosperity without growth? The transition to a sustainable economy*. Available from: http://www.sd-commission.org.uk/data/files/publications/prosperity_without_growth_report.pdf [Accessed 9 September 2012].
Szreter, S. and Woolcock M., 2004. Health by association? Social capital, social theory, and the political economy of public health. *International Journal of Epidemiology*, 33, 650–667.
Tapsell, S. M., Penning-Rowsell, E. C., Tunstall, S. M., and Wilson, T. L., 2002. Vulnerability to flooding: health and social dimensions. *Philosophical Transactions of the Royal Society A*, 360, 1511–1525.
Tebes, J. K., 2005. Community science, philosophy of science, and the practice of research. *American Journal of Community Psychology*, 35 (34), 213–30.
Thatcher, M., 1987. Margaret Thatcher, interview by Douglas Keay for *Woman's Own*, 31 October 1987. London: Thatcher Archive, COI transcript. Available from: http://www.margaretthatcher.org/speeches/displaydocument.asp?docid = 106689 [Accessed 3 August 2012].
The Copenhagen Accord, 2009. Available from: http://unfccc.int/files/meetings/cop_15/application/pdf/cop15_cph_auv.pdf [Accessed 4 September 2012].
The Copenhagen Diagnosis, 2009. *Updating the world on the latest climate science*. The University of New South Wales Climate Change Research Centre (CCRC), Sydney, Australia, 60 pp.
The Sphere Project, 2011. Humanitarian Charter and Minimum Standards in Humanitarian Response. Available from: http://www.sphereproject.org/resources/download-publications/?search = 1&keywords = &language = English&category = 22&subcat-22 = 23&subcat-29 = 0&subcat-31 = 0&subcat-35 = 0/ [Accessed 16 August 2012].
The Sphere Project, u.d. The Sphere Project in Brief. Available from: http://www.sphereproject.org/about/ [Accessed 15 August 2012].
The Widening Circle (TWC). Available from: http://www.wideningcircle.org/index.htm
Tobin, G. A. and Montz, B. E., 1997. *Natural hazards*. London: The Guildford Press.
Tol, R. S. J., 2007. Europe's long-term climate target: a critical evaluation. *Energy Policy*, 35, 424–32. doi:10.1016/j.enpol.2005.12.003.

Turner II, B. L., Kasperson, R. E., Matson, P. A., McCarthy, J. J., Corell, R. W., Christensen, L., Eckley, N., Kasperson, J. X., Luers, A., Martello, M. L., Polsky, C., Pulsipher, A., and Schiller, A., 2003. A framework for vulnerability analysis in sustainability science. *Proceedings of the National Academy of Sciences*, 100 (14), 8074–8079.

Twigg, J., 2006. Hyogo and other indicator frameworks: convergence and gaps. Note to the DFID DRR Interagency Working Group 'Creation of Community Based Indicators for Hyogo Framework of Action' project. Available from: http://www.proventionconsortium.org/themes/default/pdfs/characteristics/HFA_and_other_frameworks.pdf [Accessed 22 August 2012].

UK Cabinet Office, 2011. *Keeping the country running: natural hazards and infrastructure: a guide to improving the resilience of critical infrastructure and essential services.* Available from: http://www.cabinetoffice.gov.uk/sites/default/files/resources/natural-hazards-infrastructure.pdf

United Nations (UN), 1987. Report of the World Commission on Environment and Development: our common future. New York, NY: United Nations. Available from: http://upload.wikimedia.org/wikisource/en/d/d7/Our-common-future.pdf [Accessed 10 September 2012].

United Nations (UN), 1991. Strengthening of the coordination of humanitarian emergency assistance of the United Nations. Available from: http://www.un.org/documents/ga/res/46/a46r182.htm

United Nations Development Group (UNDG), 2007. Human rights in the context of UN reform. Available from: http://www.powershow.com/view/ac8b6-ZDJiN/Human_Rights_in_the_context_of_UN_Reform_powerpoint_ppt_presentation [Accessed 6 September 2012].

United Nations Development Programme (UNDP), 2007. *Human Development Report 2007/08.* New York, NY: Palgrave McMillan.

United Nations Framework Convention for Climate Change (UNFCCC), 1992. United Nations Framework Convention on Climate Change. Available from: http://unfccc.int/resource/docs/convkp/conveng.pdf [Accessed 11 September 2012].

United Nations Framework Convention for Climate Change (UNFCCC), 2002a. Decision 7, COP 7. Available from: http://unfccc.int/resource/docs/cop7/13a01.pdf#page=43 [Accessed 12 September 2012].

United Nations Framework Convention for Climate Change (UNFCCC), 2002b. Decision 10, COP 7. Available from: http://unfccc.int/resource/docs/cop7/13a01.pdf#page=52 [Accessed 10 September 2012].

United Nations Framework Convention on Climate Change (UNFCCC), 2007a. The Bali Road Map. Available from: http://unfccc.int/meetings/bali_dec_2007/meeting/6319/php/view/decisions.php [Accessed 5 September 2012].

United Nations Framework Convention for Climate Change (UNFCCC), 2007b. Investment and financial flows to address climate change. Bonn: Climate Change Secretariat.

United Nations Framework Convention for Climate Change (UNFCCC), 2010. Press release: *UNFCCC Executive Secretary: Governments make progress towards deciding shape of result at UN Climate Change Conference in Mexico, but need to narrow down number of negotiating options.* 6 August 2010. Available from: http://unfccc.int/files/press/news_room/press_releases_and_advisories/application/pdf/pr_20100608_closing_awg_aug.pdf [Accessed 10 September 2012].

United Nations Framework Convention for Climate Change (UNFCCC), 2011. Draft decision-/CP. 17: establishment of an ad hoc working group on the Durban platform for enhanced action. Available from: http://unfccc.int/files/meetings/durban_

nov_2011/decisions/application/pdf/cop17_durbanplatform.pdf [Accessed 12 September 2012].
United Nations General Assembly, 1986. Declaration on the Right to Development 41/128. Available from: http://www.un.org/documents/ga/res/41/a41r128.htm [Accessed 12 September 2012].
United Nations International Strategy for Disaster Reduction (UNISDR), 2004. *Living with risk: a global review of disaster reduction initiatives*. New York, NY: UNISDR Secretariat. ISBN/ISSN: 9211010640.
United Nations International Strategy for Disaster Reduction (UNISDR), 2005. Hyogo Framework for Action 2005–2015: building the resilience of nations and communities to disaster. Available from: http://www.unisdr.org/files/ 1037_hyogoframeworkforactionenglish.pdf [Accessed 14 August 2012].
United Nations International Strategy for Disaster Reduction (UNISDR), 2007. Hyogo Framework for Action 2005–2015: building the resilience of nations and communities to disasters. Extract from the final report of the World Conference on Disaster Reduction (A/CONF.206/6). Available from: http://www.unisdr.org/files/ 1037_hyogoframeworkforactionenglish.pdf [Accessed 14 August 2012].
United Nations International Strategy for Disaster Reduction (UNISDR), 2009. Terminology on disaster risk reduction. Geneva: UNISDR. Available from: http://www.unisdr.org/files/7817_UNISDRTerminologyEnglish.pdf [Accessed 14 August 2012].
United Nations International Strategy for Disaster Reduction (UNISDR), 2010. Making cities resilient: 'My city is getting ready!' Geneva: UNISDR. Available from: http://www.unisdr.org/campaign/resilientcities/about [Accessed 15 August 2012].
United Nations Office for the Coordination of Humanitarian Affairs (UNOCHA), 2012. Asean Regional Forum Inter-Sessional Meeting on Disaster Relief. Briefing by Oliver Lacey-Hall. Brisbane, Australia, 17 April 2012.
United Nations Office of the High Commissioner for Human Rights (UNOHCHR), 2006. Principles and guidelines for a human rights approach to poverty reduction strategies. Geneva: UNOHCHR.
Urry, J., 2011. *Climate change and society*. Cambridge, UK: Polity Press.
Vidal, J., 2009. Copenhagen climate summit in disarray after 'Danish text' leak, *The Guardian*, 8 December 2009. Available from: http://www.guardian.co.uk/environment/2009/dec/08/copenhagen-climate-summit-disarray-danish-text [Accessed 4 September 2012].
Vidal, J., 2010. Climate change talks yield small chance of global treaty, *The Guardian*, 11 April 2010. Available from: http://www.guardian.co.uk/environment/2010/apr/11/climate-change-talks-deal-treaty [Accessed 14 September 2012].
Vogel, C. and O'Brien, K., 2004. *Vulnerability and global environmental change: rhetoric and reality*. Information Bulletin on Global Environmental Change and Human Security. Ottowa, ON: AVISO.
Voyer, J. P. and Franke, S., 2006. Social capital as a public policy tool: conclusions from the PRI project. Presentation to the CERF Conference: Understanding cultural diversity and the economics of social inclusion and participation, 25 May 2006. Available from: http://www.cerforum.org/conferences/200605/papers/Jean-Pierre%20Voyer%20-%20CERF%20Presentation.pdf [Accessed 14 September 2012].
Walker, B., Holling, C. S., Carpenter, S. R., and Kinzig, A., 2004. Resilience, adaptability and transformability in social–ecological systems. *Ecology and Society* [online], 9 (2), 5. Available from: http://www.ecologyandsociety.org/vol9/iss2/art5 [Accessed 14 September 2012].

Wallis, J., Killerby, P., and Dollery, B., 2004. Social economics and social capital. *International Journal of Social Economics*, 31 (3), 239–258.

Ward, B. and Dubos, R., 1972. *Only one earth: the care and maintenance of a small planet.* New York, NY: W. W. Norton.

Watts, J., 2010. Climate tensions resurface as US clashes with China. *The Sydney Morning Herald*, 8 October 2010. Available from: http://www.smh.com.au/environment/climate-change/climate-tensions-resurface-as-us-clashes-with-china-20101007-169uh.html [Accessed 14 September 2012].

Weichselgartner, J., 2001. Disaster mitigation: the concept of vulnerability revisited. *Disaster Prevention and Management*, 10 (2), 85–94.

Wenger, E., 2000. Communities of practice and social learning systems. *Organization*, 7 (2) 225–246.

Wenger, E., 2004. Knowledge management as a doughnut: shaping your knowledge strategy through communities of practice. *Ivey Business Journal*, 63 (3), 1–8.

Werrity, A., Houston, D., Ball, T., Tavendale, A., and Black, A., 2007. *Exploring the social impacts of flood risk and flooding in Scotland.* Dundee, UK: University of Dundee. Available from: http://www.scotland.gov.uk/Resource/Doc/174676/0048938.pdf [Accessed 1 October 2012].

While, A., 2011. The carbon calculus and transitions in urban politics and political theory. *In*: H. Bulkeley, V. Castán Broto, M. Hodson, and S. Marvin, eds. *Cities and low carbon transitions.* London: Routledge.

Wilby, R. L. and Dessai, S., 2010. Robust adaptation to climate change. *Weather*, 62 (7), 180–185.

Williams, C. C. and Millington, A. C., 2004. The diverse and contested meanings of sustainable development. *The Geographical Journal* [online], 170 (2), 99–104. Available from: http://onlinelibrary.wiley.com/doi/10.1111/j.0016-7398.2004.00111.x/pdf [Accessed 2 October 2012].

Wilson, J. and Felsted, A., 2008. Munich Re highlights climate change impact. *The Financial Times*, 30 December 2008. Available from: http://www.ft.com/cms/s/0/fa034360-d612-11dd-a9cc-000077b07658.html#axzz1H9QPfRnE

Wisner, B., 2005. Hurricane Katrina: winds of change? 5 September 2005. Available from: http://www.radixonline.org/resources/wisner-hurricane_katrina_windsofchange_revision_5-9-05.doc. See also Radix (Radical Interpretations of Disaster) at http://www.radixonline.org/index.htm [Accessed 30 August 2012].

Wisner, B., Blaikie, P., Cannon, T., and Davis, I., 2004. *At risk: natural hazards, people's vulnerability and disasters.* 2nd ed. London: Routledge.

Woolcock, M., 1998. Social capital and economic development: toward a theoretical synthesis and policy framework. *Theory and Society*, 27, 151–208.

Woolcock, M., 2010. The rise and routinization of social capital, 1988–2008. *Annual Review of Political Science*, 13 (1), 469–87.

Woolcock, M. and Narayan D., 2000. Social capital: implications for development theory, research, and policy. *The World Bank Research Observer*, 15, 225–249.

World Bank, 2010a. *The costs to developing countries of adapting to climate change: new methods and estimate.* Washington, DC: World Bank. Available from: http://siteresources.worldbank.org/INTCC/Resources/EACCReport0928Final.pdf [Accessed 4 October 2012].

World Bank, 2010b. *The economics of adaptation to climate change.* Washington, DC: World Bank. Available from: http://siteresources.worldbank.org/EXTCC/Resources/EACC_FinalSynthesisReport0803_2010.pdf [Accessed 9 October 2012].

World Bank, 2010c. *Natural hazards, unnatural disasters: the economics of effective prevention.* World Bank, United Nations, Green Press Initiative. ISBN 978-0-8213-8050-5

World Bank, 2011. *State and trends of the carbon market 2011.* Washington, DC: World Bank.

World Meteorological Organization (WMO), 2003. *Integrated floodplain management case study: Bangladesh flood management.* Geneva: WMO.

World Wildlife Fund (WWF), 2009. *WWF Expectations for the Copenhagen Climate Deal 2009.* WWF Global Climate Policy: Position Paper. Available from: http://www.wwf.se/source.php/1237614/Copenhagen%20Expectations.pdf [Accessed 10 March 2013].

Yohe, G. and Leichenko, R., 2010. Chapter 2: Adopting a risk-based approach. *Annals of the New York Academy of Sciences* [online], 1196, 29–40. Available from: http://ugec2010.ugecproject.org/images/9/9a/CCAdaptationNYC_6_Chapter2.pdf [Accessed 12 October 2012].

Young, T., 2009. China lowers expectations of Copenhagen deal. *Business Green: Sustainable Thinking*, 03 February 2009. Available from: http://www.businessgreen.com/bg/news/1807003/china-lowers-expectations-copenhagen-deal [Accessed 18 October 2012].

Zimmerman, E., 1933. *World resources and industries: a functional appraisal of the availability of agricultural and industrial resources.* New York, NY: Harper & Bros.

INDEX

Note: Tables are indicated in bold; graphs in italics.

AA (Arctic Amplification) 54
acceptable risk, levels of 68–9, *69*
Accra Climate Change Talks 38
adaptation 31, 32–7, **33**, 66, 70, 102–3; Cancún conference measures for 46–7, 48–9; to climate change 29–30, 53, 55, 146; and climate policy 68–9, *69*, 127; costs of 103, 105–7; and development planning 97–9, 100–4, 108–9, 114–17, *115*, 148–51, 159; of humans 2–3, 31; and risk reduction 4–5, 6, 9, 162, 187–8; and social learning 173, 175, 178–9; from vulnerability 95, 96–7, 113–14; widening the stakeholders for climate risk 64, 65; *see also* climate change; disaster management; resilience building; vulnerability
adaptation continuum, the *95*, 95–6, *99*, 100–1, **101**
Adaptation Fund, the 37
adaptive capacity 109–12, *110*, **111–12**, 122, *132*; and resilience 97–102, **101**; strategies 95–7, 112–17, *113*, *114*, *115*
adaptive cycles 122–8, *123*, *124*, *126*
African disaster relief 86–7, 154
'all-hazards' approach to disaster management 75, 77, 79, 90, 133; *see also* hazards and environmental risk
ALNAP (Active Learning Network for Accountability and Performance) 156–7
AOSIS (Alliance of Small Island States) 43, 44
AWGs (Ad Hoc Working Groups) under Kyoto Protocol 36–7, 38, 43, 48

Bangkok Climate Change Talks 37
Bangladesh Cyclone Preparedness Programme 84
BAP (Bali Action Plan) 36–7
BAPA (Buenos Aires Plan of Action) 32
Barbarisation scenario of change 60
basins of attraction and resilience 120–2, *121*, *122*
Berkeley Earth Group 56
biomass 15, **15**, 20, *22*, 138
Blair, Tony 87, 164
Boer, Yvo de 43
Bolivian adaptation strategies 96, 100
bonding social capital 166, 169
Bonn Climate Change Talks 37
Bonn talks, June 2010 43–4
bottom-up programmes for adaptation 84, 97, 98, 99–100, 113, *136*, 137
bounce-back ability of resilience 2, 88, 97–8, **101**, **183**, *187*
bounce-forward ability of resilience 29, *138*, 138–9, 183, **183**, 187
boundary interactions between communities and learning 177–8
Bretton Woods conference, the 27, 148
bridging social capital 166
BRM (Bali Road Map), the 36–7
Brundtland Commission (WCED: World Commission on Environment and Development) 11–12, 139, 150–1
BSE outbreak in the UK 78

Cancún climate change conference 43, 44–9
Cap and Trade 21, 25–6, 34
capitalism 8, 28, 64, 148, 183, 184–5

Index **213**

carbon-based energy 1
carbon emissions *17*, *18*; *see also* GHGs (greenhouse gases)
carbon tax 25
carbon trading 26
cascading faults in the electricity system 23
CBDM (community-based disaster management) 84–5
CDM (Clean Development Mechanism) levy 34
change and future scenarios 59–60, 64–5, 93, *115*
chaos in a networked world 63
Charles G. Koch Charitable Foundation 56
China 14, 38, 41, 44, 149; energy use 15–16, 53
cities and resilience 15, 144–5
civic society 42–3, 60
Civil Contingencies Act 2004 80–2
climate change 5–6, 10–11, 51, 55–9, 137, 161, 182; adaptation to 29–30, 53, 55, 146; consequences of 32, 49, 53–4, 59, 63–6, 70–1; and human activity 3–4, 16, 25, 39–40, 54–5, 90; as produced hazard 133, **133**
climate change policy 68–9, *69*, 127; sustainability at local level 70, 84, 138
climate change projections 4, 39–40, 52, *52*
Climate Convention, the 31, 32
climate deniers 55–6, 58
ClimateGate e-mail scandal 39, 127
climate models and local impact scenarios 112–13
climate science 60–1
climate variability 85, **90**, 99, 103–4; *see also* climate change; extreme weather events
common wealth entitlement 10
communications and social networks 63
'communities of practice' 176–7
community groups 176–7
community resilience 170
computer models of emission trends 93
Comte, Auguste 86
conferences on sustainable development 151; *see also* post-Kyoto negotiations
consensus on climate change 3–4, 5, 37, 56, 127, 179
Conventional World scenario of incremental change 60
Copenhagen Accord on climate change 38–42, 43, 44
Copenhagen Diagnosis, The (report) 39–40, 108

COP (Conference of the Parties) to Kyoto Protocol 16–17, **19**, 32, 34, 35–8
costs of adaptation 103, 105–7

Danish text, the, leaking of 41
Darwin, Charles 1
decision-making on adaptation strategies 114–16
deforestation 36, 47, 53, 122–3; *see also* land-use forests
developed world, the 60, 110, 146; disaster management in 75, 76, 77–82, 133; energy use 10, 15, 19, 20; lifestyles in 10, 27; population issues 27, 149–50; and sustainable developemnt 140, 141–2
developing world 21, 41, 43, 45–6; and adaptation 32–5, **33**, 36, 49, 110; disaster management in 76, 84–5; energy use 10, 15, 19–20; *see also* humanitarian aid
development planning and adaptation 97–8, 102, 108–9, 114–17, *115*, 148–51, 159
DFID (Department for International Development) 88, 156; sustainable livelihoods framework *103*, 103–4
disaster management 67–8, 72–6, *75*, 82–90, *89*, 93, 154; in the developed world 77–82, **79**, **80**; and disaster risk 90–3, **91–2**; management cycle *159*, 159–60, *160*, 180, *180*; prevention instead of managing 76–7, 86, 135; and resilience 131–9, *132*, **133**, *138*, 145–6; and risk reduction 83–4, 113, 160–1, *180*, 180–1; *see also* DRR (disaster risk reduction); pre-disaster planning
disturbances (events) 119, 120–2, 125, 138
DRR (disaster risk reduction) 2, 70, 77, 88, 97, 110, 180–1; and adaptation 4–5, 6, 9, 162, 187–8; and disaster management 83–4, 113, 144–5, 161–2, 187–8; *see also* disaster management
Durban Platform for Enhanced Action 48–9

Earth Summit, Rio de Janeiro 151
ecological limits 64
ecological sustainability **12**, 12–13, 14
ecology research 119–20
economic and financial crisis of 2008 52–3
economic factors and climate change 57, 64
economic growth models 148–9, 152
ecosystem and planet life 127

electrical grid systems 20, 142–3
elite cues as influence on climate change perception 57
emergency management systems 133–4, 143
emergency powers in the UK 80–2, 133
emission rates 39, 52–3; at Copenhagen Accord 41, 42; of GHGs 16, *17*, *18*, 48; reduction of 3–4, 20, 25–6, 40, 45
end-use localised energy efficiency 21–3, *23*
energy and new resources 14–15, 19
energy corporations 26, 28, 58
energy demand **15**, 15–16, *16*, 18–19, 28
energy infrastructure 20, 21–2, 105, 142–3
energy ladder, the, and new fuels 21, *22*
energy poverty 20–1; *see also* poverty alleviation
energy systems 20–4, *23*, 24, *25*; marketisation of 62
entitlements and soft human rights 149
environmental management and the free market 25, 141
environmental movement, the 150
environmental studies 150–1
environment and sustainable development 12
Espinosa, Patricia 45
ETS (EU Emission Trading System) 26
EU, the 25, 48, 62, 159
exposure and disaster risk 92
extreme weather events 4, 6, 49, 53–4, 57–8, 67–8, 162, 182

FAO (Food and Agricultural Organisation) 87
FEMA (Federal Emergency Management Agency) 82, 133
Figueres, Christiana 44, 48
financial services institutions 8, 25, 26, 28
flooding in the UK 80, **80**, 133–4, 143
floods and landslides in Korea 67–8
flood warning systems 133, 143–4
fossil fuel use 18, 127, 138; increasing scarcity 105, 127; increasing use of 11, **15**, 15–16, 53; *see also* hydrocarbon reserves
free market, the 25–6, 28, 184; approaches to environmental management 25, 141
funding mechanisms for adaptation 32–6, **33**, 37, 104–5; resources by developed countries 42, 44, 46, 49; *see also* humanitarian aid
funding of climate denial campaign by energy sector 58

Gaia theory 90, 118, 186
GATT (General Agreement on tariffs and Trade) 27–8
GCF (Green Climate Fund) 46, 49
GCM (Global Citizens Movement) 60
GEF (Global Environment Facility) funding 34
GEMS (Global Environmental Monitoring System) 150
geographical approach to disaster research 73–5, *74*
Germany and alternative energy 14–15, 20
GHG concentration (stabilisation target) 3–4, 31, 48
GHGs (greenhouse gases) 3, 9, 11, 16, 18–19, 127
global drivers and local responses 2
globalisation 27–8, 116
global society, interconnectedness of 63
global warming 3, 9, 54, 150; *see also* climate change
gold standard, the 28
governance 115, 116, 117, 186–7; of climate risk 64, 65, 70; framework for resilience building 135–6, *136*
governments and policymakers 38, 70–1, 140; avoiding difficult decisions 25, 28, 52, 186; and disaster management 72, 75, 77, 78, 81–2, 84–5, 133; status quo thinking 58, 93
Great Transitions scenario of change 60

Habitat (UN Human Settlements Programme) 150
Hansen, James 25, 54–5
hazards and environmental risk 88–90, *89*, *90*; *see also* 'all-hazards' approach to disaster management
hazards research; *see* geographical approach to disaster research
healthcare in the UK 185
Hedegaard, Connie 44
household, the, and energy use 21–2
human activity and climate change 3–4, 16, 25, 39–40, 53–5, 90, 184
human-environment systems 120
human evolution 184
human interaction with the environment 73–4, *74*, 88, 89, 118–20, 137, 184
humanitarian aid 85–8, 154, 155–9
humanitarianism 86
human rights and development 152–4
Hurricane Katrina 76, 81, 84, 85

hydrocarbon reserves 29, 62–3
Hyogo Declaration, the 128, 135, 160–1

IDNDR (International Decade for Natural Disaster Reduction) 76
IMF, the 27–8, 149
indigenous resources, use of 20, 21–2, 23, 62, 142
individualism 78, 136
Industrial Revolution, the 26–7, 78, 116, 186
informal networks for learning 178
infrastructure, climate-proofing of 83, 105, 106, **111**, 142–3
INGOs (international non-governmental organisations) 86, 87, 155–9
innovation, belief in 64
interventions and disaster relief 87–8, 154–60, *159*, *160*
IPCC (Intergovernmental Panel on Climate Change) 5, 25, 39, 49, 177, 178–9; and adaptation 69–70, 102–3, 108; climate predictions 3, 93

JEEAR (Joint Evaluation of Emergency Assistance to Rwanda) 156–7
Johannesburg Summit, the 151

Kyoto Protocol, the 14, 16–17, 21, 31, 34, 48

land-use forests 42; *see also* deforestation
learning environments and social resilience 98, 173–81, *174*, *180*, *181*, 185
LIBOR (London Inter-Bank Rate) scandal 28
linking social capital 166
local government 68, 75, 85; and pre-disaster planning 144, 145
local level and sustainable climate change policy 70, 84, 112–13, 138; *see also* CBDM (community-based disaster management)
local wars and intervention 155
London bomb attacks 80
low carbon economy, the 14–15, **15**, 25, 48, 116, 181, *181*
'low-regret' measures to combat climate change 51, 66, 114

Mann, Michael 56
Marrakesh Accords, the 34
McKibben, Bill 58

MDGs (Millennium Development Goals) 5, 8, 15, 19–20, 103, 149, 153
media, the 6, 57, 58, 63, 76, **111**
melting of ice sheets and glaciers 4, 40
methane 4, 54, 93
methodologies of adaptation projects 96
militaristic approaches to disaster management 134
mitigation 52, 68–9, *69*, 178, 179; and adaptation 34, 35, 37, 38, 105, 117, 184; for developing countries 44, 45–6; of new technologies 9, 14, 17–18, 20, 26; and reducing GHG emissions 31, 47, 48, 104–5, 127; sustainable hazards approach of 75, *75*, 135
modernity and individualism 78, 136
Monckton of Brenchley, Viscount 57
MOP (Meeting of the Parties) of the Kyoto Protocol 17, 34
MRV (measurement, reporting and verification) of mitigation actions 45–6
multipolar world, emergence of 42

Nairobi Work Programme, the 49
NAMAs (Nationally Approriate Mitigation Actions) 45
NAPAs (National Adaptation Programmes of Action) 34
NAPs (National Adaptation Plans) 49
nation states 8–9, 42, 60, 87, 115–16
NCAP (Netherlands Climate Assistance Programme) 94, 96, 100, 101, 102
neo-liberalism 8, 14, 28, 58, 64, 149, 150
NEPIs (New Environmental Policy Instruments) 141
non-linearity of events 119
'normal science' 60–1, 64; *see also* PNS (post-normal science)
Norway 17
nuclear energy **15**, 25, 62–3, 132

OCHA (Office for the Coordination of Humanitarian Assistance) 87
OECD (Organisation for Economic Co-operation and Development) 13–14, 15, *16*, 27, 42, 140
oil producing countries and adaptation 35
organisational learning *174*, 174–8, 176

panarchy 125–6
paradigm shifts 60–1, **98**, 98–100
people and nature 14

perceptions of climate change 55–9
PNS (post-normal science) 55, 61, *61*, 64, 65; and risk management 65, 66
political consensus 3–4, 5, 135
political cycle, shortness of 52
population growth forecast 15, 26–7, 149–50
post-Kyoto negotiations 36–42, 43–50
post-normal risk management 66–70, *67, 69, 107*, 107–9
post-normal science 61, *61*, 64
poverty alleviation 94–5, 98, 103, 149–50
pre-disaster planning 96–7, 106, 135, 144–5, 154–5, *180; see also* disaster management
preparatory meetings for Cancún climate conference 43–4
'produced unknowns' of climate variability 3, 92, 99, 135, 180, *180*
production and the free market 184
punctuated equilibrium 64, 183, 186

Rasmussen, Lars Løkke 41
Red Cross, the 86, 87, 154, 157
REDD (Reducing Emissions from Deforestation and Forest Degradation) 42, 47
reflexivity and social groups 116, 136
reform and sustainable development 140
refugees, problem of 86
renewable energy 14–16, **15**, 20, 21, 23, 26, 28–9, 105
Republican party, the 56
resilience, meaning of 2, 8–10, 29–30, 118–22, 128–31, **129–31**; and adaptive cycles 122–8, *123, 124, 126*
resilience building 7, 81–2, *187*, 187–8; and adaptation 95–6, *96*, 97–100, **98, 99**, 146; barriers to 85, 147; and basins of attraction 120–2, *121, 122*; and disaster management 131–9, *132*, **133**, *138*, 145–6, *180*, 180–1; from the household up 6, 21–5, *23, 24*; schools of **183**, 183–4; and social capital 147, 170; and sustainable developemnt *139*, 139–44
Rio +20 summit 182
risk management 5–6, 52–60, 69, 70, 77, **79**; and PNS 65, 66; *see also* disaster management; post-normal risk management
risk reduction and disaster management; *see* DRR (disaster risk reduction)
Rostow, W.W. 148

RRF (Regional Resilience Forums) 81
Rwandan crisis, the, and humanitarian aid 155–6

SBSTA (Subsidiary Body for Scientific and Technological Advice) 35, 37, 43
scale of adaptation processes 96
SCAT (Social Capital Assessment Tool) 166–7, *167*, 172
SCCF (Special Climate Change Fund), the 34, 35
scenario planning 59–60, 93, 112–13, *115*; *see also* development planning and adaptation
science 60–1, 64
scientific realism 64
sea levels 40
seepage of methane in the Arctic 93
Sen, Amartya 152
'shadow states' 60
social capital 163, 164–70, *165, 167, 168*; measurement of 171–3, *172*
social change since Industrial Revolution 115–16
social classes 185
social learning 173–81, *174, 180, 181*
sociological approach to disaster research 73
Sphere Project, the 157, *158*
stability landscape and 'basin of attraction' 121, *121*
stakeholders 61–2, 102, *107*, 176, 178–9, 181, *181*; and governance of climate risk 64, 65, 70
state planning 148–9
status quo thinking 25, 58–9, **79**, 110; and climate negotiations 14, 28–9, 41, 69; and economic growth 15, 141
Stern, Lord 52
structural and non-structural measures of risk reduction 83
subsidies to energy sector 26
Sun, the 11
supply-side issues in energy systems 20
support for climate change 39
sustainability and economics 1, 182–3
sustainable development 5, 51, 109, 135, 151; meaning of 11–14, **12**, *13*; and resilience 128, *136, 139*, 139–44
sustainable hazards approach of mitigation 75, *75*, 135

technocratic model of disaster management **79**
technologically driven environmental management 141
technological resilience 142–3
technology transfer to the developing world 21, 47, 70
temperature increases 4, 39–40, 52, *52*; limiting global increases 43, 45, 53
terrorism 155, 175; and risk 79–80, 81, 82
Thames Estuary 2100 Plan 68–9, *69*
tipping points 16, 40, 53, 93, 108, 185–6
top-down structures 9, 90, 98, **98**, 99–100, **183**; and disaster management 79, **79**, 133–5, *136*, 138, 147; electrical systems 23, *24*, 142–3
transformation and sustainable development 141
transition period of world today 62
tsunami disaster in Japan 83
TWC (The Widening Circle) 60

Ujamaa, Tanzanian development policy 148–9, 150
UKCIP (UK Climate Impact Programme) 108
UK disaster management 79–82, **80**
UN, the 87, 152–3
UNCED (UN Conference on Environment and Development) 151
uncertainty in science 61–2
uncertainty of climate change 55, 59, 62, 63, 93; and adaptation 53, 65, 109, *114*
UNDP (United Nations Development Programme) 5, 152

UNEP (United Nations Environment Programme) 150, 162
UNFCCC (UN Framework Convention for Climate Change) 5, 7, 42–4, 178; funding for adaptation 32–4, **33**, 49, 102–3, 106; objectives and GHGs 11, 16–17, 127, 162
UNFPA (United Nations Population Fund) 150
UNHCR (United Nations High Commissioner for Refugees) 86, 87
UNICEF (UN Children Fund) 87
UNISDR (UN International Strategy for Disaster Reduction) 5, 7, 72; building resilience 76–7, 83, 128, 144
urbanisation 144, 145, 149–50

vulnerability 104, 109–10, 135, 142, 163–4; and adaptation 100, 113–14, 131; and disaster risk **91–2**, 92–3, 94, 95, 145, 170–1

Washington Consensus, the 28, 42, 149
wealth creation 8
weather trends 54–5
Western humanitarian interventions 87–8, 155–6
WFP (World Food Programme) 87
'wicked problems' 55, 65, 90, 182
World Bank, the 27–8, 46, 106, 148
WSSD (World Summit on Sustainable Development) 128

Yokohama Strategy, the 76